Intelligent Computing

Ruizhuo Song • Qinglai Wei
Qing Li • Shi Xing

Intelligent Computing

Concepts, Principles and Applications

Ruizhuo Song
School of Automation
University of Science and
Technology Beijing
Beijing, China

Qing Li
School of Automation
University of Science and
Technology Beijing
Beijing, China

Qinglai Wei
Institute of Automation
Beijing, China

Shi Xing
University of Science and
Technology Beijing
Beijing, China

ISBN 978-981-96-9789-2   ISBN 978-981-96-9790-8 (eBook)
https://doi.org/10.1007/978-981-96-9790-8

Jointly published with Tsinghua University Press, Beijing, P.R.China

The print edition is not for sale in Mainland China. Customers from Mainland China please order the print book from: Tsinghua University Press.

© Tsinghua University Press 2026

This work is subject to copyright. All rights are solely and exclusively licensed by the Publisher, whether the whole or part of the material is concerned, specifically the rights of reprinting, reuse of illustrations, recitation, broadcasting, reproduction on microfilms or in any other physical way, and transmission or information storage and retrieval, electronic adaptation, computer software, or by similar or dissimilar methodology now known or hereafter developed.
The use of general descriptive names, registered names, trademarks, service marks, etc. in this publication does not imply, even in the absence of a specific statement, that such names are exempt from the relevant protective laws and regulations and therefore free for general use.
The publishers, the authors, and the editors are safe to assume that the advice and information in this book are believed to be true and accurate at the date of publication. Neither the publishers nor the authors or the editors give a warranty, express or implied, with respect to the material contained herein or for any errors or omissions that may have been made. The publishers remain neutral with regard to jurisdictional claims in published maps and institutional affiliations.

This Springer imprint is published by the registered company Springer Nature Singapore Pte Ltd.
The registered company address is: 152 Beach Road, #21-01/04 Gateway East, Singapore 189721, Singapore

If disposing of this product, please recycle the paper.

Preface

Intelligent computing is a general term for computing technologies developed from traditional computing methods and different from traditional computing methods. Intelligent computing is a computing method developed by humans inspired by the objective laws of natural biological groups and biological thinking, movement and other behaviors, including evolutionary computing, swarm intelligence computing, neural computing, and many other algorithms. Most of these algorithms realize the purpose of intelligent computing by simulating the special functions of some species in the nature or some characteristics of the nature, and program and execute the wisdom of biological groups and some natural laws, so as to design optimization algorithms with intelligent nature.

Therefore, the essence of intelligent computing can be understood as the method and technology of solving complex problems by simulating biological and natural intelligence, abstracting specific problems into mathematical models for description, and on this basis, comprehensively using programming, computing, visualization, and other technologies to conduct knowledge mining and rule sorting on the data therein.

With the rapid development of computer technology, the era of big data has arrived, which also enables AI to rise and develop rapidly. Machine learning is the core content of artificial intelligence research, and also the fundamental way to enable computers to achieve intelligent computing. At present, machine learning has become an important branch and new development direction of intelligent computing.

This book studies the core computing methods that make up intelligent computing, takes classical evolutionary computing as the touchpoint, and extends to swarm intelligence computing and neural computing algorithms based on the inspiration of natural biological movement behavior, providing the latest research results. At the same time, with the rapid development of artificial intelligence, this book also makes a detailed study and introduction of different algorithm classifications and application examples of machine learning, the latest development direction of intelligent computing.

Considering the wide application of intelligent computing algorithms at present, this book adds a large number of algorithm model examples and corresponding Python or MATLAB code and introduces the latest research trends of corresponding algorithms in each chapter, which is convenient for readers to operate and reproduce. It is suggested that readers further expand and think about the calculation examples provided in this book.

This book is the joint effort of the editor and his team. Thank the team members HuaFeng Zhang, Ying Zhou, Kexin Wen, and Shi Xing. In the process of compiling this book, the above students participated in and helped to complete the data collection, code sorting, and implementation, as well as the peer MD work of manuscript proofreading.

Beijing, China

Ruizhuo Song
Qinglai Wei
Qing Li
Shi Xing

Contents

1 Introduction.. 1
  1.1 Overview of Intelligent Computing.......................... 1
  1.2 Evolutionary Calculations.................................. 2
  1.3 Group Intelligent Computing................................ 2
  1.4 Neural Computation... 3
  References.. 4

2 Genetic Algorithms in Evolutionary Computation................ 7
  2.1 Overview of the GA... 7
    2.1.1 Genetic Algorithm.................................... 7
    2.1.2 Basic Schematic Diagram.............................. 7
    2.1.3 Pattern Theorem...................................... 8
    2.1.4 Building Block Hypothesis............................ 8
    2.1.5 Research Progress.................................... 9
  2.2 Process Flow of the Genetic Algorithm...................... 9
    2.2.1 Scientific Definition................................ 9
    2.2.2 Execution Process.................................... 10
    2.2.3 Basic Essence.. 10
    2.2.4 Chromosomal Coding................................... 11
    2.2.5 Population Initialization............................ 11
    2.2.6 Evaluation of the Fitness Value...................... 11
    2.2.7 Select the Operator.................................. 12
    2.2.8 Cross-Over Operator.................................. 12
    2.2.9 Variant Operator..................................... 13
    2.2.10 Flow Chart and Pseudocode........................... 14
  2.3 Improvement of the Genetic Algorithm....................... 14
    2.3.1 Operator Selection................................... 14
    2.3.2 Parameter Setting.................................... 15
    2.3.3 Hybrid Genetic Algorithm............................. 16
    2.3.4 Parallel Genetic Algorithm........................... 17

	2.4	Encoding Rules of the Genetic Algorithm.	18
		2.4.1 Binary Encoding Method.	18
		2.4.2 Floating-Point Encoding Method.	19
		2.4.3 Symbol Encoding Method	19
	2.5	Application of the Genetic Algorithm	20
	2.6	Related Application of Genetic Algorithm and MATLAB Example.	21
		2.6.1 Genetic Algorithm Example 1	21
		2.6.2 Genetic Algorithm Example 2	23
	2.7	Summary of the GA.	27
	Reference		27
3	**Group Intelligent Computing**		29
	3.1	Particle Swarm Optimization.	29
		3.1.1 Introduction to the Particle Swarm Optimization Algorithm.	29
		3.1.2 Basic Flow of the Particle Swarm Optimization Algorithm.	33
		3.1.3 Classification of the Particle Group Algorithm	35
		3.1.4 Study on the Improvement of the Particle Swarm Optimization Algorithm.	39
		3.1.5 Parameter Settings of the Particle Swarm Optimization Algorithm.	41
		3.1.6 Comparison of the Particle Swarm Optimization Algorithm and the Genetic Algorithm	43
		3.1.7 Related Applications of Particle Swarm Optimization Algorithm and MATLAB Examples	44
	3.2	Ant Colony Algorithm	49
		3.2.1 Basic Principles of the Ant Colony Algorithm	49
		3.2.2 Algorithm Process of the Ant Colony Algorithm	51
		3.2.3 Development of the Ant Colony Algorithm.	54
		3.2.4 Improvement of the Ant Colony Algorithm.	56
		3.2.5 Parameter Setting of the Ant Colony Algorithm	59
		3.2.6 Application of the Ant Colony Algorithm	59
		3.2.7 Related Application of Ant Colony Algorithm and MATLAB Example.	61
	References.		65
4	**Neural Computing**		67
	4.1	BP Neural Network	67
		4.1.1 Concept of the BP Neural Network.	67
		4.1.2 Model of the BP Neural Network	67
		4.1.3 Characteristics of the BP Neural Network.	76
		4.1.4 Related Applications of BP Neural Network and MATLAB Examples	78
		4.1.5 Algorithmic Improvement of the BP Neural Network.	79

	4.2	Deep Neural Network	88
		4.2.1 Concept of the Deep Neural Network	88
		4.2.2 Model of Deep Neural Network	89
		4.2.3 Characteristics of Deep Neural Networks	99
		4.2.4 Applications of Deep Neural Networks	101
		4.2.5 Optimization of Deep Neural Networks	102
	4.3	Convolutional Neural Networks	103
		4.3.1 History and Basic Concepts of Convolutional Neural Networks	103
		4.3.2 Structure of Convolutional Neural Network	105
		4.3.3 Application of Convolutional Neural Network and MATLAB Example	110
		4.3.4 Latest Development of Convolutional Neural Network	113
	4.4	Recurrent Neural Network	128
		4.4.1 Structure of Recurrent Neural Network	128
		4.4.2 Application of Recurrent Neural Network and MATLAB Examples	129
		4.4.3 Latest Development of Recurrent Neural Networks	134
	References		150
5	**Machine Learning**		151
	5.1	Naive Bayes Algorithm	151
		5.1.1 The Basic Concept of the Naive Bayes Algorithm	151
		5.1.2 The Process and Model of the Naive Bayes Algorithm	154
		5.1.3 Characteristics and Application Scenarios of the Naive Bayes Algorithm	155
		5.1.4 Relevant Applications of the Naive Bayes Algorithm and MATLAB Examples	156
	5.2	Decision Trees	165
		5.2.1 The Basic Concept of Decision Trees	166
		5.2.2 Construction of Decision Trees	168
		5.2.3 Pruning of Decision Trees	173
		5.2.4 Implementation of Decision Tree Algorithms	174
		5.2.5 Relevant Applications of Decision Trees and MATLAB Calculation Examples	182
	5.3	Random Forest	205
		5.3.1 The Basic Concept of Random Forest	206
		5.3.2 Construction Method of Random Forests	209
		5.3.3 Extension of Random Forests	211
		5.3.4 Relevant Applications of Random Forests and MATLAB Examples	217
	References		225

Chapter 1
Introduction

1.1 Overview of Intelligent Computing

Intelligent computing (IC) is a general term for computing techniques developed based on traditional computing methods and different from traditional computing methods. Inspired by the objective laws of biological groups and biological thinking and movement in nature, intelligent computing has developed computing methods in many algorithmic fields such as intelligent computing, including evolutionary computing, group intelligent computing, neural computing and so on. Most of these algorithms realize the purpose of intelligent computing by simulating the special functions of some species in nature or some characteristics of nature, and program and execute the wisdom of biological groups and some natural laws of biological groups, so as to design the optimization algorithm with intelligent nature.

Therefore, the essence of intelligent computing can be understood as the methods and techniques to solve complex problems by simulating biological and natural intelligence, abstracting specific problems into mathematical models, and on this basis, comprehensively using programming, computing and visualization technologies to conduct knowledge mining and law sorting of the data.

With the rapid development of computer technology, the era of big data has arrived, which also enables the rapid rise and development of artificial intelligence. Machine learning (ML) is the core content of artificial intelligence research, and it is also the fundamental way to achieve intelligent computing by using computers. At present, machine learning has also become an important branch and a new development direction of intelligent computing.

Intelligent computing is the product of the cross-evolution of ancient disciplines such as mathematics, physics, psychology, physiology, neuroscience and emerging disciplines such as life science, cognitive science and computer science. Due to the wide application range of intelligent computing and the powerful computing and optimization capabilities, it has achieved large-scale applications in the fields such as solving the optimization problems in complex engineering or social application background.

The following mainly starts from the four main branches of intelligent computing, introduces the generation and development process of intelligent computing, and summarizes the current development trend of intelligent computing.

1.2 Evolutionary Calculations

Evolutionary computing is a very important branch in the field of intelligent computing. Its basic idea is to solve optimization problems in actual production and life by simulating the natural evolution process, which conforms to the rule of "survival of the fittest" in nature.

In 1948, the British mathematician Alan Turing first proposed the algorithmic idea of evolutionary search. Further, in the late 1950s, biologists proposed the concept and basic idea of the evolutionary inheritance of the genetic algorithm (GA), which was systematically proposed by Professor Holland in 1962 (Holland 1975). A year later, German mathematicians I. Rechenberg and H. P. Schwefel proposed that natural selection was performed in a definite manner with only one evolutionary strategy algorithm. In 1965, American scientist Fogel proposed the Evolutionary Planning algorithm, which is an algorithm to solve the parameter optimization problem by imitating the principle of natural evolution (Fogel 1994). The principle is similar to the evolutionary strategy, but emphasizes the group-level behavior changes in natural evolution and is suitable for solving the problem of objective function. By the 1990s, with the genetic programming ideas, evolutionary calculation as a subject is formally proposed and accepted by the world, and the combination with other intelligent computing method of theory and applied research also got the attention of institutions and scholars, also derived such as Memetic algorithm (global search and local search mixed algorithm), differential evolution (DE) algorithm, group of wisdom Energy Swarm Intelligence calculation algorithm and other improved methods. At present, evolutionary computing has become an important research direction in the field of artificial intelligence in intelligent computing.

1.3 Group Intelligent Computing

Based on the wide application and continuous development of evolutionary computing, scientists continue their attention to the evolution of the biological world. Many social organisms (including ant colonies, birds, fish, etc.) work together to complete many complex behaviors that individual behaviors cannot complete. With the deepening of research, scientists have designed the biomimetic optimization algorithm based on the self-organization behavior of social organisms through abstract simulation, which has been widely practiced and applied in many fields.

In 1960, American scientist Steele proposed the concept of bionics, which is to mimic or a systems science with and similar to the characteristics of biological systems. Later, in the research field of optimization algorithm, the first algorithm with the biological evolution process as the simulation object appeared, such as the evolutionary strategy algorithm mentioned in Sect. 1.2. After nearly 30 years of development, in the 1990s, scientists put forward a variety of group intelligent computing algorithms with perfect theories and remarkable effects, including ant colony algorithm, particle swarm algorithm, fish swarm algorithm, cat swarm algorithm and frog jumping algorithm, etc., group intelligent computing has gradually become an independent science.

The ant colony algorithm is a probabilistic algorithm used to find optimized paths. It was proposed by Marco Dorigo in his doctoral thesis in 1992, and was inspired by the study of mechanisms coordinating ants searching for food paths and communication in colony foraging (Martinoli 2001). This algorithm is characterized by distributed collaboration, information positive feedback and heuristic search, and is essentially a heuristic global optimization algorithm in evolutionary algorithms. By conducting parallel computing on multiple individuals at the same time, the computing power and operation efficiency of the algorithm are greatly improved, but also has the characteristics of not easy to fall into the local optimum, easy to find the global optimal solution.

Particle swarm algorithm is a random search algorithm based on group collaboration developed by simulating bird flock foraging behavior. It was proposed by American psychologist by Dr. Kennedy and electrical scientist Dr. Eberhart (Kennedy and Eberhart 1995). The inspiration comes from the simulation of predation behavior of the flock, where the optimal strategy for find food is by searching the surrounding area of the bird closest to the food.

Group intelligent computing has the advantages of simple algorithm principle, less parameter setting, good optimization effect and fast use, which has stimulated peoples enthusiasm for continuous research and innovation, and a variety of new ideas and ideas are constantly emerging. In a word, group intelligent computing, as the most novel and cutting-edge intelligent computing method, has a very broad development prospect. At the same time, it also needs to test the application value and prospect of its latest expanded research through continuous verification in practical problems.

1.4 Neural Computation

Since the invention of computers, humans have been trying to use machines instead of the human brain for the calculation and reasoning of complex systems. Therefore, computer-based intelligent computing methods emerge in large numbers and develop continuously, resulting in artificial neural network algorithms based on neural computing.

As early as 1890, psychologist William James proposed the conjecture that neural cell activation is the result of the superposition of all cell inputs, which established the concept for the proposal of artificial neural network. At the beginning of the nineteenth century, the Italian anatomist Golgi confirmed that the cranial nerve was composed of a variety of cells independent and with clear boundaries, also known as the concept of "neuron".

In 1943, psychologist McCulloch and logician Pitts put forward the mathematical description and structure of neurons, and established the logic calculus function of artificial neurons mathematical model, its pioneering work for artificial neural network junction, structural research laid a solid foundation, and developed more neural network model, such as perceptron and adaptive linear components. However, in the late 1960s, Minsky, a famous artificial intelligence expert in the United States, published the book Perceptron, which pointed out the limitations of the function and processing capacity of the *perceptron* (Minsky and Papert 1969), and did not give a calculation method that can improve the processing ability of the neuronal network, which made the research of neural network enter a low tide.

In 1982, Professor Hopfield of California Institute of Technology proposed a new neural network model Hopfield network for the first time (Hopfield 2000). By introducing the concept of "energy function", he gave the clear criterion of neural network stability, and applied it to the classic problem of optimization field—NP-hard, and achieved very good results. In 1986, Professor Rumelhart and McCelland proposed a multi-layer feedforward network with error back propagation, namely BP neural network (BackPropagationNeuralNetwork), based on the multi-layer neural network model, which has become one of the most successful and widely used artificial neural network models. Since then, the development of neural network has entered a period of rapid development and practical application.

At present, artificial neural network algorithms based on neural computing are widely used in pattern recognition, signal processing, knowledge engineering, decision assistance, optimization combination, robot control and other fields in production and life. With the great evolution of computer computing power and the vigorous development of big data technology, multi-layer neural networks with deep learning functions represented by deep neural network and convolutional neural network have pushed scientists' enthusiasm for neural network research to a new height. In the future, the development of neural network will be more extensive and climb to a new scientific peak.

References

D.B. Fogel, An Introduction to Simulated Evolutionary Optimization, EEE Trans. on NN,Vol.5, No.1, Jan., 1994, pp3-14.

J. H. Holland, Adaptation In Natural and Artificial Systems, University of Michigan, 1975.

Hopfield J J. On theorists and data in computational neuroscience. Nature Neuroscience, 2000, 3 (Suppl 11):1204-1204.

References

J. Kennedy and R. Eberhart, Particle swarm optimization, Proceedings of ICNN'95 - International Conference on Neural Networks, Perth, WA, Australia, 1995, pp. 1942-1948 vol.4, doi: https://doi.org/10.1109/ICNN.1995.488968.

Martinoli A. Collective Complexity out of Individual Simplicity: A Review of Swarm Intelligence: From Natural to Artificial Systems, by Eric Bonabeau, Marco Dorigo, and Guy Theraulaz. Artificial Life, 2001, 7(3):315-319.

M.L. Minsky and S.A. Papert, Perceptrons: An introduction to computational geometry (MIT Press, Cambridge, 1969).

Chapter 2
Genetic Algorithms in Evolutionary Computation

2.1 Overview of the GA

2.1.1 Genetic Algorithm

Genetic algorithm is a branch of evolutionary computation, a stochastic search algorithm that simulates the evolutionary process of organisms in nature.

The idea of GA originates from the evolutionary law of "natural selection" and "survival of the fittest" in nature, and seeks the global optimal solution of the problem by simulating natural selection and mating variation in biological evolution. It was first proposed by John H. Holland, a university professor in Michigan, and has now been widely used in optimization problems in various engineering fields.

2.1.2 Basic Schematic Diagram

Figure 2.1 describes the rationale of the genetic algorithm.

There is an analogy between traditional biological genetic evolution processes and genetic algorithms.

1. Biological genetic evolution: the evolution of mating variation in chromosomal gene adaptation ability in population populations ends.
2. Genetic algorithm: a set of effective solutions in the search space to obtain a new population feasible solution encoding a string chromosome of a coding unit of the chromosome exchange part of the end of the numerical change algorithm.

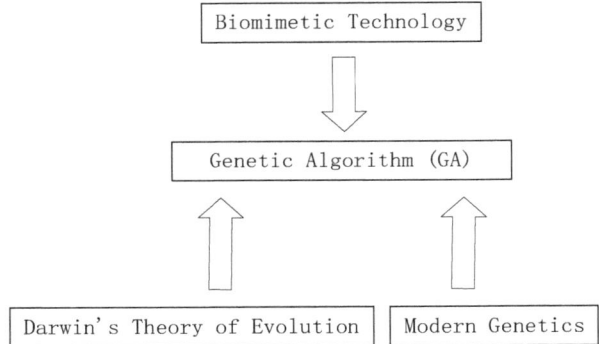

Fig. 2.1 Basic schematic diagram

2.1.3 Pattern Theorem

Holland The pattern theorem mainly involves the following three concepts.

1. Mode: the collection of chromosomes with similar structure at certain positions encoded in the population.
2. Order of the mode: refers to the number of genes with determined values in the mode.
3. The definition length of a pattern: refers to the distance from the first gene with a certain value to the last gene with a certain value in the pattern.

Holland Model theorem proposed, the essence of the genetic algorithm is through selection, crossover and variation operator to search patterns, low order, define the length is small and the average fitness value is higher than the group average fitness value in the proportion of the population will increase exponentially, as the evolution, the number of better chromosomes will increase rapidly.

2.1.4 Building Block Hypothesis

Building blocks are patterns of low-order, small defined length, and with average fitness values higher than the population average fitness value.

The block hypothesis is that in the process of genetic algorithm operation, blocks can be combined with each other under the influence of genetic operators to produce new and better blocks, and eventually close to the global optimal solution.

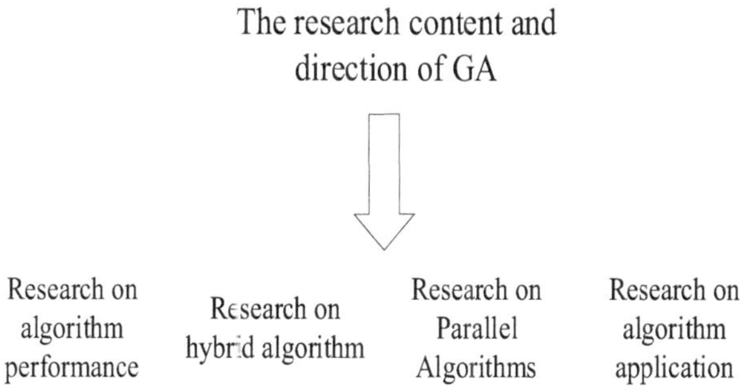

Fig. 2.2 GA Research direction

The research content and direction of GA
⇩

- Research on algorithm performance
- Research on hybrid algorithm
- Research on Parallel Algorithms
- Research on algorithm application

2.1.5 Research Progress

At present, the research focus of genetic algorithm focuses on algorithm performance, mixed algorithm, parallel algorithm and algorithm application. Figure 2.2 describes the direction of research on the genetic algorithm.

2.2 Process Flow of the Genetic Algorithm

2.2.1 Scientific Definition

Genetic algorithm is a computational model of biological evolutionary processes that simulate the natural selection and genetics mechanism of Darwinian biological evolution, and is a method to search for optimal solutions by simulating natural evolutionary processes. The advantages of the GA are shown as follows.

1. Directly operate on the structure object, and there is no limit of derivative and function continuity.
2. Have the inherent hidden parallelism and the better global optimization search ability.
3. Using the probabilistic optimization method, it can automatically obtain and guide the optimized search space without certain rules, and adjust the search direction adaptively.

Genetic algorithms take all individuals in a population as objects and use randomization technology to efficiently search a coded parameter space. Among them, selection, crossover and mutation constitute the genetic operation of genetic algorithms; parameter encoding, initial population setting, fitness function design, genetic operation design, and control parameter setting constitute the core content of genetic algorithms.

2.2.2 Execution Process

The genetic algorithm starts with a population that represents the potential solution set of the problem, and a population consists of a certain number of individuals encoded by a gene. Each individual is actually an entity with a characteristic chromosome. As the main carrier of genetic material, that is, the collection of multiple genes, its internal expression (i.e., genotype) is a kind of combination of genes, which determines the external expression of individual shape. For example, the characteristics of black hair are determined by a combination of genes that control this feature in the chromosome. Therefore, it is necessary to implement the first place. Because of the complexity of imitating gene coding, it needs to simplify using binary coding.

After the emergence of the initial population, according to the principle of survival of the fittest and survival of the fittest, the generation evolution produces better and better approximate solutions. At each generation, individuals are selected according to the size of the fitness of the individual in the problem domain, and crossover and variation are combined with the help of the natural genetic operator to produce a population representing the new solution set.

This process will lead to the natural evolution of the population being more adapted to the environment than the previous generation, and the optimal individual in the last population is decoded, which can be used as the approximate optimal solution to the problem.

2.2.3 Basic Essence

Genetic algorithms are a class of robust search algorithms that can be used to optimize complex systems. As a fast, simple and fault-tolerant algorithm, the genetic algorithm shows obvious advantages in the optimization process of all kinds of structured objects. Compared with the traditional optimization algorithm, the GA mainly has the following features.

1. The search process does not directly act on the variables, but on the coded individuals. This encoding operation enables genetic algorithms to operate directly on structural objects (sets, sequences, matrices, trees, graphs, chains, and tables).
2. The search process is iterated from one group of solutions to another. The method of processing multiple individuals in the group at the same time, which reduces the possibility of falling into the local optimal solution and is easy to be parallelized.
3. Use the probability change rules to guide the search direction, rather than use the deterministic search rules. There are no special requirements for the search space (such as connectivity, convexity, etc.), only using adaptive information, no other auxiliary information such as derivatives, and a wider range of adaptation.
4. The genetic algorithm directly uses the fitness as the search information, without other auxiliary information such as derivatives. The genetic algorithm uses the

2.2 Process Flow of the Genetic Algorithm

search information of multiple points, which has implicit parallelism. Genetic algorithms use probabilistic search techniques rather than deterministic rules.

Genetic algorithm takes the encoding of decision variables as the operation object.

2.2.4 Chromosomal Coding

At present, several commonly used encoding technologies include binary encoding, floating point encoding, character encoding, programming encoding, etc. Binary encoding is the most common encoding method in genetic algorithms, that is, the binary character set {0, 1} generates a common string of 0 and 1 to represent the candidate solution of the problem.

For example, assuming that the decimal number between [U_{min}, U_{max}] is encoded in binary, the encoding method shown in Fig. 2.3 can be used.

2.2.5 Population Initialization

In general, the genetic algorithm adopts the random number initialization method in the population initialization stage, and adopts the method of generating random numbers to initialize the value of each dimensional variable of the chromosome. The initializing the chromosome must pay attention to whether the chromosome satisfies the definition of the effective solution of the optimization problem.

If the initial population is guaranteed to be already excellent to a certain extent at the beginning of evolution, it will effectively improve the ability of the algorithm to find the global optimal solution.

2.2.6 Evaluation of the Fitness Value

The evaluation function is used to evaluate the fitness value of each chromosome, and then to distinguish the good and the bad. The evaluation function is often determined according to the optimization goal of the problem, for example, when solving

$$U_{min} \longrightarrow 0000\cdots0000$$

$$U_{max} \longrightarrow 1111\cdots1111$$

$$U_{min} + \frac{(U_{max}-U_{min})\sum_{j=1}^{l.} X_j 2^{j-1}}{2^l - 1} \longrightarrow X_l X_{l-1} \cdots X_2 X_1$$

Fig. 2.3 The chromosome encoding Fig

the function optimization problem, the objective function defined by the problem can be used as the prototype of the evaluation function.

In GA, chromosomes with larger prescribed fitness values are better. Therefore, for some numerical optimization problems that solve the maximum value, we can directly apply the function expressions defined by the problem. However, for other optimization problems, the objective function expression of the problem definition must be somewhat transformed.

2.2.7 Select the Operator

Selection operators commonly used in genetic algorithms include roulette selection, binary tournaments, linear ordering, and exponential ordering. The basic genetic algorithm generally adopts wheel selection, and Fig. 2.4 shows the process of wheel selection.

The roulette selection operator first values the sum of the adaptation values of all chromosomes in the population according to the adaptation of each chromosome in the population, and calculates Pi for the sum of the adaptation values and the population adaptation values separately. Secondly, assume a roulette with N sectors, each sector corresponds to a chromosome in the population, and the size of the sector is proportional to the Pi value of the corresponding chromosome.

2.2.8 Cross-Over Operator

In the chromosome cross stage, each chromosome can mate is determined by cross probability Pc (general value of 0.4 ~ 0.99), the specific process is: for each chromosome, if Random (0,1) less than Pc, Random (0,1) is the uniform distribution of

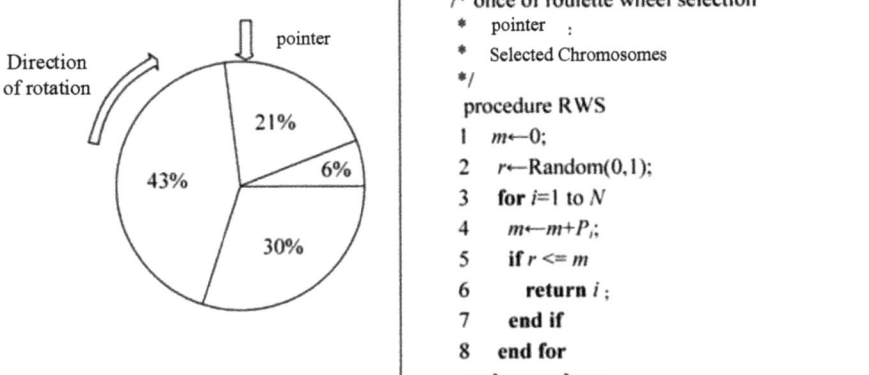

Fig. 2.4 The wheel selection diagram

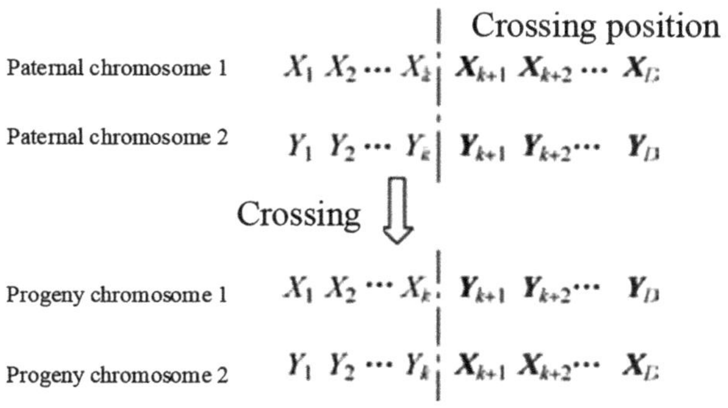

2.2 Process Flow of the Genetic Algorithm

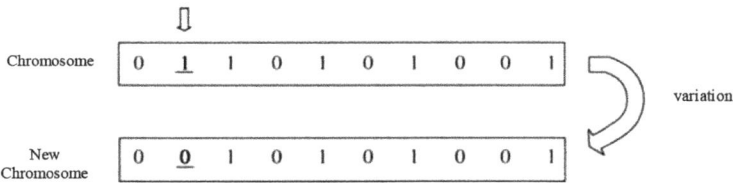

Fig. 2.5 Cross Fig

Fig. 2.6 Variant Fig

random number between [0,1], means the chromosome can cross operation, otherwise the chromosome does not participate in the cross directly copied into the new population.

Each two chromosomes selected according to the Pc crossover probability are crossed, after the exchange of the respective partial genes, to produce two new progeny chromosomes. The specific operation is to randomly generate an effective crossover position after the chromosome exchange is located after all genes. Figure 2.5 depicts the chromosome crossover map.

2.2.9 Variant Operator

The variation of the chromosome acts on the gene. For each position of the chromosome in the new population after crossing, whether the gene is determined according to the mutation probability Pm.

Random (0,1) is the random number evenly distributed among [0,1]. If Random (0,1) is less than Pm, the value of the gene is changed, otherwise the gene does not vary and remains unchanged. Figure 2.6 depicts the chromosomal variants.

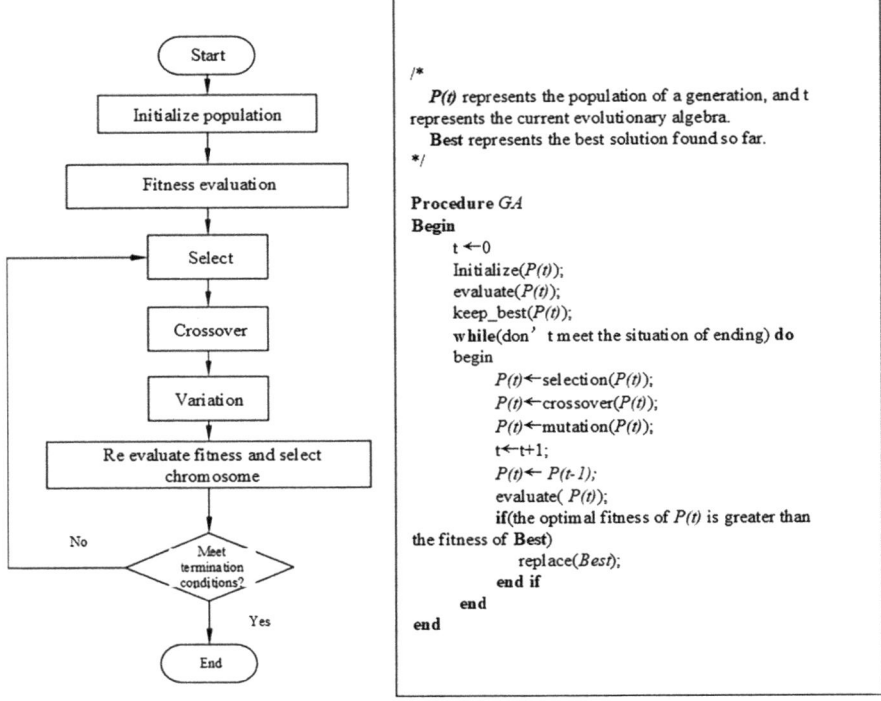

Fig. 2.7 Flow and code of chromosomes from selection to variation

2.2.10 Flow Chart and Pseudocode

Figure 2.7 depicts the process of chromosome from selection to variation and the algorithm process of selection.

2.3 Improvement of the Genetic Algorithm

2.3.1 Operator Selection

The improvement of genetic algorithm is mainly reflected in the operator selection, and the operators of genetic algorithm mainly include the following types.

1. Selection operator: adaptive value proportional model, best individual preservation model, exclusion model, expected value model, stochastic tournament model, ranking model, deterministic sampling, no playback remainder random sampling.
2. Crossover operator: multi-point crossover operator, partial matching crossover operator, sequential crossover operator, cyclic crossover operator, edge recombi-

nation crossover operator, edge set crossover operator, two-point crossover operator, uniform crossover operator, arithmetic crossover operator, monosexual spore crossover operator.
3. Variation operator: boundary variation operator, Gaussian variation operator, and non-uniform variation operator.

2.3.2 Parameter Setting

1. Group size N: It affects the search ability and operation efficiency of the algorithm. If N is set in a large size and there are more modes covered by one evolution, the diversity of the group can be guaranteed, so that the search ability of the algorithm can be improved. However, due to the large number of chromosomes in the group, the calculation amount of the algorithm will be increased and the operating efficiency of the algorithm will be reduced. If the N set is small, it reduces the computation, but also reduces the ability to contain more and better chromosomes in each evolution. N is generally set at values ranging from 20 to 100.
2. Chromosome length L: it affects the calculation amount of the algorithm and the effect of mating variation operation. The setting of L is closely related to the optimization problem, which is generally determined by the form of the solution defined by the problem and the coding method chosen. For the binary encoding method, the length L of the chromosome selects the size based on the value range of the solution and the specified accuracy requirements. For the floating point encoding method, the length L of the chromosome is the same as the dimension D of the solution defined by the problem. Basically, the chromosome length will follow a certain coding method. Goldberg et al. also proposed a variable length chromosome genetic algorithm, Messy GA, where the length of the chromosome is not fixed.
3. Gene value range R: Determined by the chromosome encoding scheme used. For binary encoding method, R = {0,1}, while for floating point encoding method, R is the same as the value range of the solution to the optimization problem.
4. Crossing probability Pc: determines the average number of chromosomes in which the population attended the evolutionary process. The value is generally 0.4 ~ 0.99. Adaptive method can also be used to adjust the crossover probability during algorithm operation.
5. Variation probability Pm: increasing the diversity of population evolution determines the average number of genes that the population changes in the process of evolution. The value of Pm should not be too large. Because the variation has a certain destructive effect on the found better solution, if the value of Pm is too large, the current good search state of the algorithm may go back to the original poor situation. The value of Pm is generally 0.001 ~ 0.1, and the adaptive method can also be used to adjust the Pm value during the operation of the algorithm.
6. Fitness value evaluation: influence the population selection of the algorithm. The appropriate evaluation function should be able to make an appropriate distinc-

tion between the advantages and disadvantages of chromosomes, ensure the effectiveness of the selection mechanism, so as to improve the evolutionary ability of the group. The setting of the evaluation function is related to the solving objective of the optimization problem. The evaluation function should satisfy the requirement of larger adaptation values for the better chromosomes. In order to better improve the performance of the selection, some corrections can be made to the evaluation function. At present, the main evaluation function correction methods are linear transformation, multiplied power transformation and exponential transformation.
7. Termination condition: determines when the algorithm stops running and outputs the optimal solution found. The termination condition used is related to the application of the specific problem. The termination condition can make the algorithm stop when the maximum evolutionary number is reached. The maximum evolutionary number can generally be set to 100 ~ 1000. The recommended value can be modified accordingly according to the specific problem; the algorithm can also be stopped by examining whether the current optimal solution found meets the error requirement; or the algorithm stops when the optimal solution found by the algorithm for a long period of evolution has not been improved.

2.3.3 Hybrid Genetic Algorithm

Hybrid genetic algorithm is a very efficient optimization algorithm that utilizes genetic algorithm and other efficient optimization techniques to search for high-quality optimal solutions. Therefore, hybrid genetic algorithms can be used to optimize complex practical problems, especially those that cannot be optimized directly using standard genetic algorithms. In recent years, hybrid genetic algorithms have received increasing attention in optimization theory.

The hybrid genetic algorithm is a hybrid optimization algorithm that combines two different optimization algorithms, so it enables flexibility from one optimization algorithm and powerful search performance from another. Two typical hybrid genetic algorithms are hybrid algorithms between genetic and simulated annealing algorithms, and between genetic and particle swarm algorithms.

Although the genetic algorithm has good search performance, it also has some disadvantages, such as low convergence rate, the result is easy to fall into the local optimal solution, it is too dependent on randomness, and therefore poor convergence. However, the simulated annealing algorithm has good convergence performance, but it also has its own shortcomings, such as the limit of the search performance is poor, easy to fall into a stagnation state, especially in the face of non-convex optimization problems, it has some limitations in solving the problem.

To overcome these shortcomings, the emergence of a hybrid genetic algorithm is a good option. In general, the workflow of hybrid genetic algorithms is as follows. First, the genetic algorithm is used to generate a set of initial solutions from the original solution space; second, to form the new generation of solutions according

2.3 Improvement of the Genetic Algorithm

to the given evolution strategy; finally, the other optimization algorithm leads to a very effective optimal solution.

Hybrid genetic algorithms have great applications in optimization theory. It can solve complex practical problems, such as pattern recognition, engineering design, system optimization, machine learning, etc. Typical applications include simulation optimization, engineering design optimization, and aircraft stress and durability optimization.

2.3.4 Parallel Genetic Algorithm

There are two calculation methods of parallel genetic algorithm: decomposition parallel method and standard parallel method.

Parallel computing includes single instruction flow multi-data flow computer, multiple instruction flow multi-data flow computer and parallel computing network. Serial calculation refers to the single instruction stream and the single data stream processor. Figures 2.8 and 2.9 describe the standard parallel method network and the decomposition parallel method network, respectively.

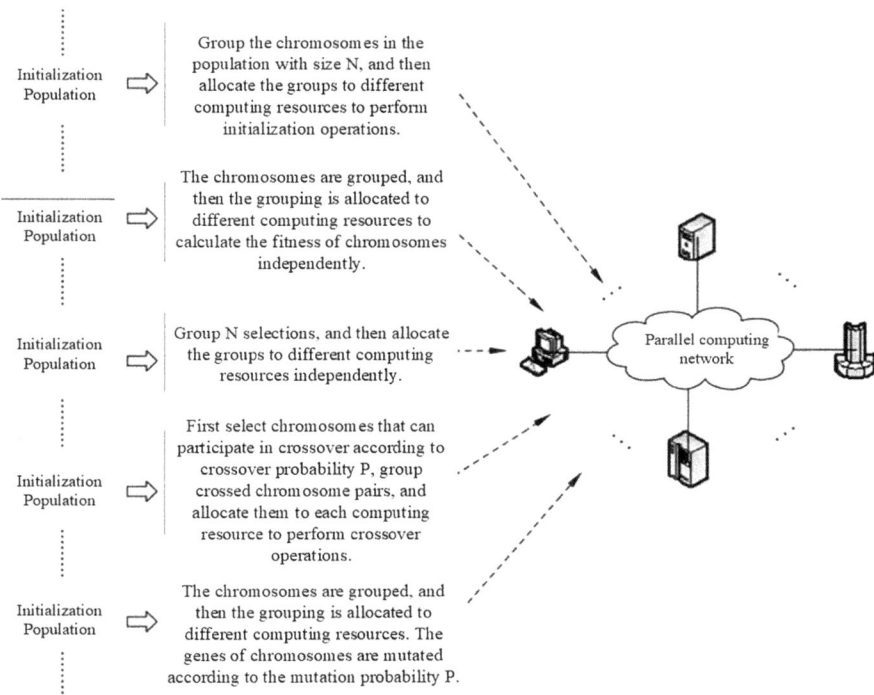

Fig. 2.8 The Standard Parallel Method network

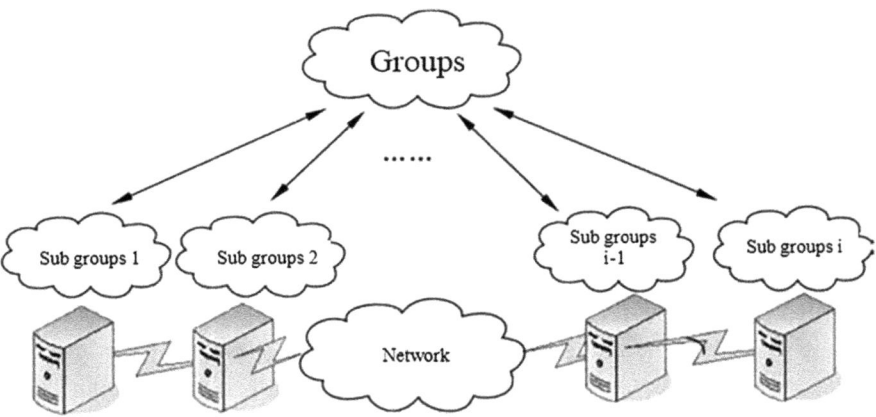

Fig. 2.9 To decompose the parallel method network

In Fig. 2.9, the following issues should be noted in the exchange of subgroup evolutionary information, namely, the time of the exchange, the way of the exchange, and the content of the exchange.

2.4 Encoding Rules of the Genetic Algorithm

Coding is the primary problem to be solved when applying genetic algorithms, and it is also a key step in designing genetic algorithms. The coding method affects the operation methods of genetic operators such as crossover operators and mutation operators, and largely determines the efficiency of genetic evolution. So far, many different coding methods have been proposed. Generally speaking, these coding methods can be divided into three categories: binary coding, floating point coding, and symbolic coding.

2.4.1 Binary Encoding Method

Just like human genes have AGCT 4 base sequences, the genetic algorithm uses two bases, 0 and 1, and then strings them into a strand to form a chromosome. One digit of a chromosome can represent the amount of information in two states, so a sufficiently long binary chromosome can represent all features, which is binary coding. Binary encoding is a set of binary symbols composed of binary symbols 0 and 1. It has the following advantages.

1. Coding and decoding are simple and easy.
2. Genetic operation such as crossover and variation are easy to realize.

2.4 Encoding Rules of the Genetic Algorithm

3. Compliance with the minimum character set coding principle.
4. The mode theorem can be used to analyze the algorithm theoretically.

The disadvantage of binary coding is that for some optimization problems of continuous functions, its local search ability is poor due to its randomness. For example, for some high-precision problems, when the solution is close to the optimal solution, the expression is very variable and discontinuous, so it will be far away from the optimal solution and cannot reach stability.

2.4.2 Floating-Point Encoding Method

Although binary coding is simple and intuitive, there has mapping error in discretization of continuous functions. When the individual length is short, it may not meet the accuracy requirements, while when the individual coding length is long, although the accuracy can be improved, it increases the difficulty of decoding, which makes the search space of genetic algorithm sharply expand.

The floating-point method means that each gene value of an individual is expressed by a floating-point number in a certain range. In the floating-point encoding method, it must be guaranteed that the gene value is within the given interval limit range, and the genetic operators such as crossover and variation used in the genetic algorithm must also ensure that the gene values of new individuals produced by its operation results are also within this interval limit range, for example, 1.2-3. 2-5.3-7.2-1.4-9.7. The floating-point number encoding method has the following advantages.

1. It is suitable to represent the numbers with a large range in the genetic algorithm.
2. It is suitable for genetic algorithms with high accuracy requirements.
3. To facilitate the genetic search in a large space.
4. Improve the computational complexity of the genetic algorithm and improve the operational efficiency.
5. To facilitate the mixed use of genetic algorithm and classical optimization methods.
6. Easy to deal with complex decision variable constraints.

2.4.3 Symbol Encoding Method

Symbolic coding means that the gene value in the individual chromosome code string is taken from a set of symbols that have no numerical meaning but only code meaning, such as {A, B, C, ...}. The main advantages of symbolic coding are as follows.

1. According to the coding principle of meaningful building blocks.
2. to facilitate the use of expertise for solving the problem in genetic algorithms.
3. To facilitate the mixed use of genetic algorithms and related approximation algorithms.

2.5 Application of the Genetic Algorithm

Because the overall search strategy and optimization of genetic algorithm search method does not rely on gradient information or other auxiliary knowledge, but only need to affect the search direction of the target function and the corresponding fitness function, so the genetic algorithm provides a general framework for solving complex system problems, it does not rely on the specific areas of the problem, the problem is robust, so widely used in many fields, will introduce some main areas of the application of genetic algorithm.

Function optimization is a classical application field of genetic algorithm, and also a common example for performance evaluation of genetic algorithm. At present, a variety of complex test functions have been constructed: continuous function and discrete function, convex and concave function, low dimensional function, high dimensional function, unimodal function and multimodal function. For some nonlinear, multi-model and multi-objective function optimization problems, it is difficult to solve by other optimization methods, and better results can be easily obtained by using the genetic algorithm.

With the increase of the problem scale, the search space of the combinatorial optimization problem also increases sharply. Sometimes, it is difficult to find the optimal solution. For such complex problems, it has been realized that people should focus on seeking satisfactory solutions, and genetic algorithms are one of the best tools for seeking such satisfactory solutions. GA has proved very effective for NP problems in combinatorial optimization. For example, genetic algorithms have been successful in solving travel business problems (TSP), backpack problems, work schedules, test group questions, packing problems, and graphical division problems.

The optimization problem can also be solved by using genetic algorithms. First, the feasible domain is encoded (generally binary coding), then some coding groups are randomly selected in the feasible domain as the first generation of coding groups from the starting point of evolution, and the objective function value of each solution, that is, the fitness of the encoding. A selection mechanism was used to randomly select codes from the coding group as the coding samples before the breeding process. The selection mechanism should ensure that solutions with higher fitness can retain more samples, while solutions with lower fitness can retain fewer samples and even be eliminated. In the following propagation process, the genetic algorithm provides the two operators of crossover and variation to exchange the selected samples. The cross operator exchanges certain bits of two randomly selected codes, and the variant operator reverses a randomly selected bit in a code, so that the next generation coding group is generated through selection and reproduction. The above selection and breeding processes were repeated until the end conditions were met. The optimal solution in the last generation of the evolutionary process is the final result of solving the optimization problem with a genetic algorithm.

2.6 Related Application of Genetic Algorithm and MATLAB Example

2.6.1 Genetic Algorithm Example 1

For the function y = sin x + x cos x, find the maximum value of the function on [0,2π]. First you need to draw the image of the function, as shown in Fig. 2.10.

Encoding

Almost all practical problems cannot be drawn when converted into functions. This is only for observation. The first step to solve the problem is encoding, and only one variable in the problem x needs to be encoded.

Assuming that the coding length is 5, the solution interval is divided into 105 copies, but note that the first part is 0, not 1, the longer the length, the higher the accuracy.

The decoding method is very simple, for example, the x corresponding to gene 12,345 is 12345 ÷ 99999 × (2π − 0) + 0 ≈ 0.7757. It is obvious that gene 00000 represents x = 0 and 99,999 represents x = 2π. The coding of continuous problems is mostly done in this way. The coding of discrete problems needs to be designed

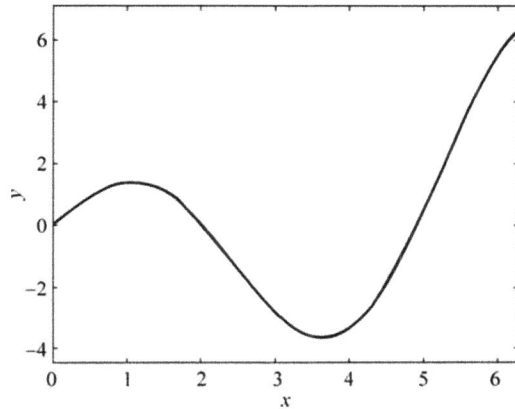

Fig. 2.10 The image of the function

according to the specific problems, and the quality of the coding design directly affects the final result of the algorithm.

The coding of discrete problems is also very common, such as TSP problems, NP problems and other decision practical problems mostly need to use discrete coding, generally not in the basic stage, so not introduced here.

Cross

Croover is a very important operation in genetic algorithm, and its advantages affect the convergence speed of the algorithm. Assuming that the genes required to cross are 12,345 and 66,666, the breakpoint randomly takes 3 and generates two new solutions from the sequence exchange after the third position: 12,366 and 66,645. The selection of breakpoints should meet the uniform distribution.

Variation

Variiation is another important operation in genetic algorithm, which affects the final result of the global optimum level. Assuming that the gene satisfying the variation condition is 66,666, when the mutation point is randomly taken 2, the value at the second position is randomly replaced to generate a new solution 60,666. The generation of this 0 and the selection of mutation points should meet the uniform distribution. It is also possible to mutate to 68,666, and the two expectations should be equal.

Natural Selection

In general, natural selection includes roulette selection methods and ranking methods.

The roulette selection method is measured by the fitness ratio of each individual and the overall ratio, the greater the less likely it is to be eliminated. Of course, the algorithm complexity of the roulette selection method is relatively high:

$$P_i = F_i \div \sum_{i=1}^{N} F_i$$

The ranking method is calculated through the fitness ranking of each individual. The higher the ranking is, the less likely it is to be eliminated. The algorithm complexity of the ranking method is relatively low:

$$P_i = R_i \div N$$

Or

2.6 Related Application of Genetic Algorithm and MATLAB Example

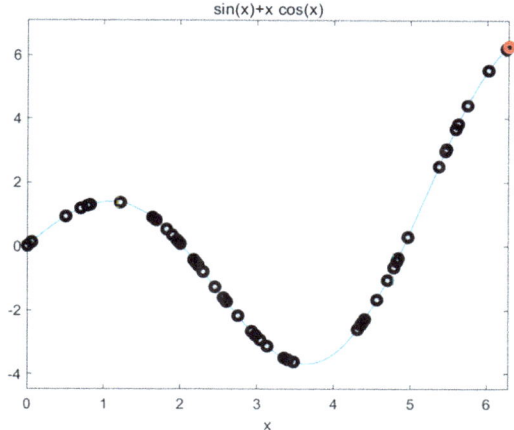

Fig. 2.11 The max value under the genetic algorithm

$$P_i = (R_i - 1) \div N$$

Where P_i is the probability that the ith individual is eliminated, F_i is the fitness of the ith individual, R_i is the fitness ranking of the ith individual, and N is the number of individuals in the population. The roulette selection method is used in the genetic prototype, but this method has many disadvantages in addition to fast calculation, so the ranking method is recommended. That is, the lower the fitness ranking is, the higher the probability of being eliminated. Figure 2.11 depicts the maximum value under the genetic algorithm.

2.6.2 Genetic Algorithm Example 2

For a practical problem of genetic algorithm to find the basic idea is: first the problem of candidate solution coding, after coding candidate solution called individual, many candidate solution coding (individual) of candidate solution group called group, selection for such groups like biological evolution, cross and variation operation, produce a new generation of groups. In each generation, the number of individuals is kept constant, and each individual is evaluated by calculating the fitness function value. The crossover and mutation operations are to ensure that the globally optimal solution is obtained. Each time the genetic algorithm completes such an operation is called a "generation". After several generations of evolution, the optimal solution to the problem can be obtain.

In bulk, the number of individuals is kept as a fixed value, and each individual is evaluated by the calculation of the fitness function value, where the crossover and variation operations are used to ensure that the solution has a global optimum. Each time a genetic algorithm completes such an operation is called "generation", after several generations of evolution, the optimal solution of the problem can be obtained.

It is customary to call the genetic algorithm proposed by Holland in 1975 as the traditional genetic algorithm. Its main steps are as follows.

1. Coding: Before the genetic algorithm, the search first represents the solution data of the solution space into the genotype string structure data of the genetic space, and the different combinations of these string structure data constitute different points.
2. Formation of the initial group: N initial string structure data are randomly generated. Each string data structure is called an individual, and N individuals constitute a group. The GA then iterates with these N string data structures as the initial points.
3. Fitness value evaluation and detection: the fitness function indicates the quality of the individual or the solution. The fitness function is defined differently for different problems.
4. Selection: Select the next generation of individuals according to the principle of survival of the fittest. In choosing, fitness is taken as the selection principle. The purpose of selection is to select excellent individuals from the current group, giving them the opportunity to reproduce as the next generation. The principle of making selection is that adaptable individuals have a large probability of contributing one or more offspring to the next generation. The selection embodies Darwin principles of survival of the fittest.
5. Crossover: Cross operation is the most important genetic operation in the genetic algorithm. A new generation of individuals can be obtained by crossover manipulation. For individuals selected for breeding the next generation, the same positions of two individuals are randomly selected and exchanged at the selected position by crossover probability. This process reflects the aim of random information exchange to generate new gene combinations, i.e., new individuals. When crossing, a single-point crossover or a multi-point crossover can be implemented.
6. Variation: variation first randomly selects an individual in the group, and randomly changes the value of a string in the selected individual data with a certain probability. According to the principle of gene variation in biological genetics, the variation is performed with the variation probability Pm for some bits of some individuals. Upon mutation, the correspondence of the string that performs the mutation is reversed, from 1 to 0 and 0 to 1. The variation probability P m is consistent with the biological variation is very small, so the value of Pm is small. Variiation cannot be benefited in the solution. However, it guarantees that the algorithmic process does not produce a single population that cannot evolve. Because in all individuals, too, crossover cannot produce new individuals, then can only rely on variation to produce new individuals. That is, variation increases the global optimization trait.
7. Global optimal convergence (convergence to the global optimum): When the fitness of the optimal individual reaches a given threshold, or the fitness of the optimal individual and the group fitness no longer rise, the iterative process of the algorithm converges and the algorithm ends. Otherwise, the previous generation group is replaced with a new generation group obtained through selection,

2.6 Related Application of Genetic Algorithm and MATLAB Example

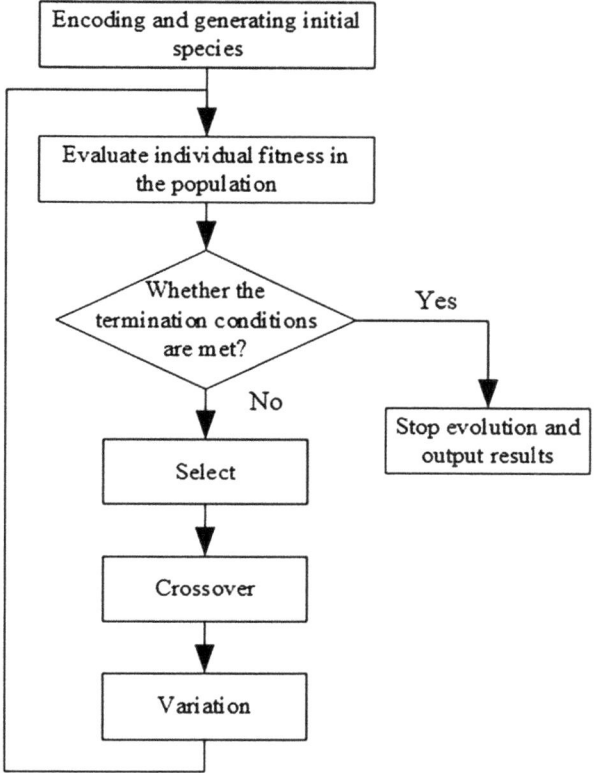

Fig. 2.12 Basic algorithm flow chart of genetic algorithm

crossover, and variation, and returns to the selection operation to continue the cycle.

The basic algorithm flow of the genetic algorithm is shown in Fig. 2.12.

The selection, crossover, and mutation in the above process are basic genetic operators, and their implementations are diverse. In recent years, different genetic gene expression patterns, crossover and mutation operators, and regeneration and selection methods have been continuously proposed, all of which can improve certain performance of GA. The encoding method of the string, determination of the fitness function, and setting of the parameters of the genetic algorithm itself are the most critical issues in the application of genetic algorithms.

Example 2.1 max $f(x) = 32x-x^2$, $x \in [0,32]$, $x \in Z$.

Solution For object X, 8 initial values are randomly generated, and 4 values are set: 6, 13, 20, 29, coded in binary:

$6 = (00110)_2$, $13 = (01101)_2$, $20 = (10100)_2$, $29 = (11101)_2$

Which

$$P_i = \frac{f(x_i)}{\frac{1}{n}\sum_{k=1}^{m} f(x_k)}$$

is the relative fitness of the individual x_i.

According to the coding of the initial population shown in Table 2.1, it can be seen that $f(29)$ has the lowest fitness and $f(13)$ has the highest fitness, so canceling $f(29)$ reproduces the coding table of the new population shown in Table 2.2.

First, the random method is used to determine the position of the cross-transposition on the string, as shown in Fig. 2.13, and then the genetic algorithm is 0 to 1 and 1 to 0.

Table 2.1 Coding table of the initial population

Individual groups	The initial group	x_i	$f(xi)$	p_i
1	00110	6	156	0.855
2	01101	13	247	1.353
3	10100	20	240	1.351
4	11101	29	87	0.476

Table 2.2 Coding tables for the new populations

Individual groups	The initial group	x_i	$f(xi)$	p_i
1	00110	6	156	0.855
2	01101	13	247	1.353
3	10100	20	240	1.351

Note: If no new individuals are generated, some fragments of the chromosome (string) can be cross-replaced according to the biological hybrid methods. Cross-transposition is pairing by random methods on paired chromosomes, for example pairing numbers 1 with numbers 2 and 3 with numbers 4

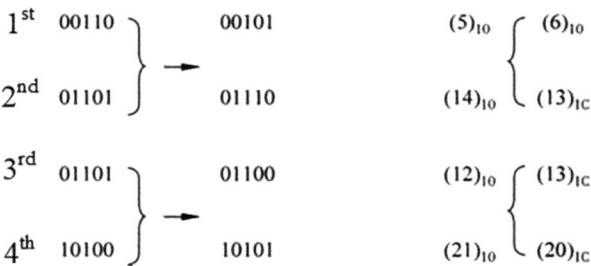

Fig. 2.13 Random method

2.7 Summary of the GA

The rise of GA research was in the late 1980s and early 1990s, but its historical origins date back to the early 1960s. Early studies have focused on computer simulations of natural systems. For example, in Frasers simulation study, he proposed concepts and ideas very similar to the current genetic algorithm. The creative research results of Holland and DeJong have changed the unobjective nature of early GA research and the lack of theoretical guidance. Among them, Holland systematically expounded the basic theory and methods of genetic algorithms in his famous 1975 book, The Adaptation of natural Systems and Artificial Systems, and put forward the model theory that is very important to the theoretical research and development of genetic algorithms. This theory confirms for the first time the importance of structural recombinant genetic manipulation for the acquisition of implicit parallelism. In the same year, De Jongs important paper "Behavior Analysis by Genetic Adaptive Systems" combined Hollands pattern theory with his computational experiments and proposed new genetic manipulation techniques such as the generation gap (De Jong 1975). It can be argued that the research work done by DeJong is a milestone in the development of genetic algorithms. In the 1980s, the genetic algorithm ushered in a prosperous period of development, and both the theoretical research and the applied research have become a very hot topic. Especially for the application of genetic algorithms. The domain is also expanding. At present, the main fields involved in genetic algorithms are automatic control, planning and design, combinatorial optimization, image processing, signal processing, and artificial life.

In the 1990s, the application research of genetic algorithm became particularly active, and its application field gradually expanded. The ability of using genetic algorithm to optimize and learn rules has also been significantly improved. At the same time, the research on industrial application is also groping. In addition, some new theories and methods have also been developed rapidly in application research, which has added new vitality to genetic algorithm. The application research of genetic algorithm has expanded from the initial combination optimization solution to many newer and more engineering applications. Genetic algorithm still has a lot of problems to be studied, and its mathematical basic theory needs to be further studied to prove its technical advantages in theory; At the same time, it is necessary to study the hardware based genetic algorithm and the general programming and form of genetic algorithm.

Reference

De Jong, K. A. (1975). An analysis of the behavior of a class of genetic adaptive systems (Doc-. (2007).

Chapter 3
Group Intelligent Computing

3.1 Particle Swarm Optimization

3.1.1 Introduction to the Particle Swarm Optimization Algorithm

Particle swarm optimization (Particle Swarm Optimization, PSO) algorithm is a branch of evolutionary computing, is a random search algorithm to simulate the biological activities of nature, also known as particle swarm algorithm, particle swarm algorithm or particle swarm optimization algorithm. The PSO algorithm is generally considered as a kind of intelligent algorithm, which can also be classified as a multi-agent optimization system (Multiagent Optimization System, MAOS).

The PSO algorithm simulates the process of bird feeding and fish feeding in nature. The global optimal solution of the problem is found through the collaboration in the group. It was proposed in 1995 by American scholars Eberhart and Kennedy and has now been widely used for optimization problems in a variety of engineering fields.

Source of Thought

The origin of PSO algorithm is based on biological phenomena and social psychology, which includes group behavior, group migration, biological foraging, etc.; social psychology includes group wisdom, individual cognition, and social influence.

PSO algorithm is Kennedy and Eberhart inspired by the results of artificial life, by simulating the process of migration and clustering behavior of a global random search algorithm based on group intelligence, all kinds of organisms in nature have certain group behavior, and one of the main research areas of artificial life is to explore the natural biological group behavior, so as to build its group model on the computer. The group behavior of birds and fish in nature has always been the research interest of scientists. The biologist Craig Reynolds proposed a very influential bird aggregation model in 1987 (Reynolds 1987). In his simulation, each individual follows:

1. Avoid collision with individual neighbors;
2. The speed of matching the neighborhood individuals;
3. Fly to the center of the flock, and the whole group flies to the target.

In the simulation, only the above three simple rules can be used to closely simulate the phenomenon of birds flying. In 1995, James Kennedy, an American social psychologist, and Russell Eberhart, an electrical engineer, jointly proposed the particle swarm optimization algorithm (Kennedy 2011). Its basic idea is inspired by the research results of modeling and simulation of bird group behavior. Their model and simulation algorithm mainly modified Frank Heppner's model, so that particles flew to the solution space and landed at the best solution.

Algorithm Background

The basic core of the PSO algorithm is to use the sharing of information between individuals in the group so that the movement of the whole group produces the evolution process from disorder to order in the problem solution space, so as to obtain the optimal solution of the problem. Imagine a scenario where a group of birds forage, and there is a cornfield in the distance, all the birds do not know where the cornfield is, but they know how far away they are from the cornfield. So the best, simplest and most effective strategy to find the cornfield is to search for the surrounding area of the birds currently closest to the cornfield.

In the PSO algorithm, the solution of each optimization problem is a bird in the search space, called the "particle", and the optimal solution of the problem corresponds to the "cornfield" sought by the birds flock. All particles have a position vector (the position of the particle in the solution space) and a velocity vector (determining the direction and speed of the next flight), and can calculate the fitness value of the current position based on the objective function, which can be understood as the distance from the "cornfield". In each iteration, the particles in the population can learn not only from their own experience (historical location), but also from the experience of the optimal particles in the population, so as to determine how the direction and speed of flight need to be adjusted and changed in the next iteration. In this way, the particles of the whole population will gradually tend to the optimal solution.

Fundamentals

Bird foraging and PSO algorithm have an analogy relationship to some extent.

1. Bird flock foraging phenomenon: birds, foraging space, flight speed, location, individual cognition and group cooperation, to find food.

3.1 Particle Swarm Optimization

2. PSO algorithm: a set of effective solutions to the search space, the search space of the problem, the speed vector of the solution, the position vector of the solution, the speed and position update, and find the global optimal solution.

Figures 3.1 and 3.2 describe the flock foraging phenomenon and the PSO algorithm, respectively.

Algorithm Definition

The PSO algorithm simulates the predation behavior of the bird flocks. A flock of birds was randomly searching for food, with only one piece of food in the area. All the birds have no idea where the food is. But they know how far away they are from the food. So whats the best strategy to find food? The simplest and most effective thing is to search for the area around the bird closest to the food.

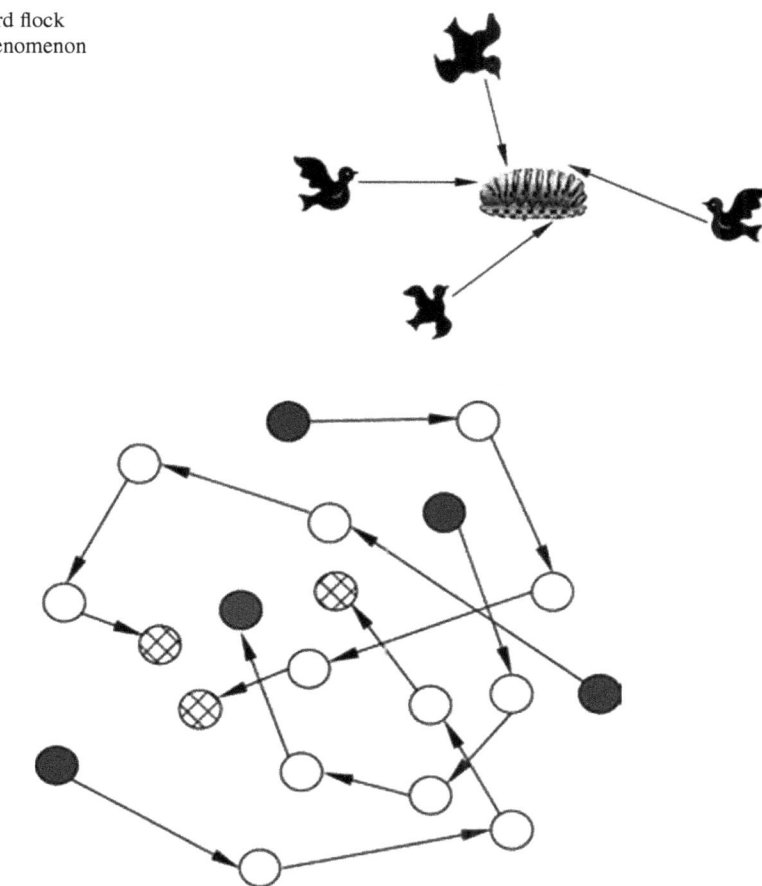

Fig. 3.1 Bird flock foraging phenomenon

Fig. 3.2 PSO algorithm

According to the model of prey behavior, the PSO algorithm is learned and used to solve the optimization problem. In the PSO algorithm, the solution of each optimization problem is a particle in the search space. All particles have a fitness value determined by the optimized function, and each particle also has a velocity that determines their orientation and distance. The particles then follow the current optimal particles to search in the solution space.

The PSO algorithm is initialized as a population of random particles (random solutions), and the optimal solution is found iteratively. In each iteration, the particles update themselves by tracking two "extreme values". The first one is the optimal solution found by the particle itself, which is called the individual extreme value pBest, the other extremum is the optimal solution found by the whole population, and this extreme value is the global extreme value gBest. In addition, you can use the neighbors of a part of the optimal particles instead of the whole population, which is the local extremum in all the neighbors.

There are two important steps in the application of the PSO algorithm to solve the optimization problem: the encoding of the problem solution and the fitness function. One advantage of PSO algorithm is the use of real coding, not binary coding like genetic algorithm.

There are no special parameters in the PSO algorithm. Common parameters and empirical settings are listed below.

1. The number of particles is generally taken as 20 ~ 40. For most of the problems, 10 particles are good enough to achieve good results. For more difficult problems or specific categories of problems, the number of particles can be taken up to 100 or 200.
2. The length of the particle is determined by the optimization problem, which is the length of the problem solution.
3. The range of the particle is determined by the optimization problem, and each dimension can set a different range.
4. V max represents the maximum velocity, which determines the maximum movement distance that a particle is in a cycle, and is usually set to the range width of the particle.
5. The learning factors c_1 and c_2 are generally valued as 2. There are also different values in other literature, such as c_1 equals c_2 and range from 0 to 4.
6. The suspension condition is the maximum number of cycles and the minimum error requirement. The suspension condition is determined by the specific problem.
7. Global PSO and local PSO, the former speed is fast but sometimes fall into the local optimum, the latter convergence rate is a little slow but it is difficult to fall into the local optima. In practice, you can find the general result with the global PSO, and then search with the local PSO.

3.1.2 Basic Flow of the Particle Swarm Optimization Algorithm

Assume that the basic conditions of particle swarm optimization algorithm are as follows:

1. The target search space is a D-dimensional space.
2. The particle population is composed of N particles.
3. The position of the ith particle in D-dimensional space is $X_i = \{x_i^1, x_i^2, x_i^3, \ldots, x_i^D\}$
4. The flying velocity of the ith particle is $V_i = \{v_i^1, v_i^2, v_i^3, \ldots, v_i^D\}$
5. The optimal position of the ith particle $pBest_i = \{pBest_i^1, pBest_i^2, pBest_i^3, \ldots, pBest_i^D\}$
6. Global optimal position $gBest_i = \{gBest_i^1, gBest_i^2, gBest_i^3, \ldots, gBest_i^D\}$ then the individual speed and position update formula of the PSO algorithm particles are respectively:

$$v_i^d = \omega \times v_i^d + c_1 \times r_1^d \times (pBest_i^d - x_i^d) + c_2 \times r_2^d \times (gBest^d - x_i^d) \quad (3.1)$$

$$x_i^d = x_i^d + v_i^d \quad (3.2)$$

Where d = 1, 2, ... , D, c_1, c_2 are non-negative constants; i = 1, 2, ... , m; r_1, r_2 are random numbers between [0,1]; $V_d^i \in [-V_{max}, V_{max}]$ is setted.

Figure 3.3 describes the overview of particle swarm optimization algorithm. The speed update formula consists of three parts.

1. Self speed: also the inertia or momentum part, which reflects the motion habits of particles.
2. Individual cognition: particles have the advantage of approaching the best position in their history.
3. With social guidance, particles tend to approach the best position in the history of groups or fields.

Fig. 3.3 Algorithm Overview

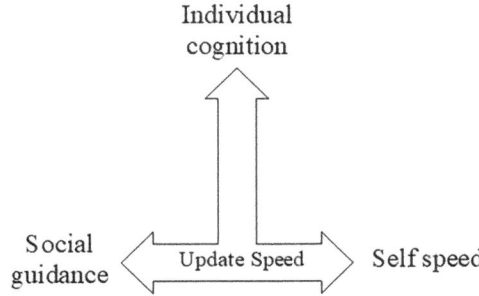

Therefore, in formula (3.1), the velocity of particle in step d own velocity inertia in the previous step and individual cognitive part and social recognition.

The knowing part, that is, the speed is the sum of three parts.

The flow and pseudo code of the particle swarm optimization algorithm are shown in Fig. 3.4.

Artificial neural network (ANN) is a simple mathematical model that simulates the process of brain analysis, and reverse broadcast algorithm is the most popular neural network training algorithm. Recently, many scholars have used evolutionary computing technology to study various aspects of artificial neural network.

Evolutionary computing generally studies three aspects of neural network: network connection weight, network structure (network topology and transfer function), and network learning algorithm. Most of the current work focuses on the network connection weights and the network topology. In genetic algorithms, the network weight and/or topology are generally encoded as chromosomes

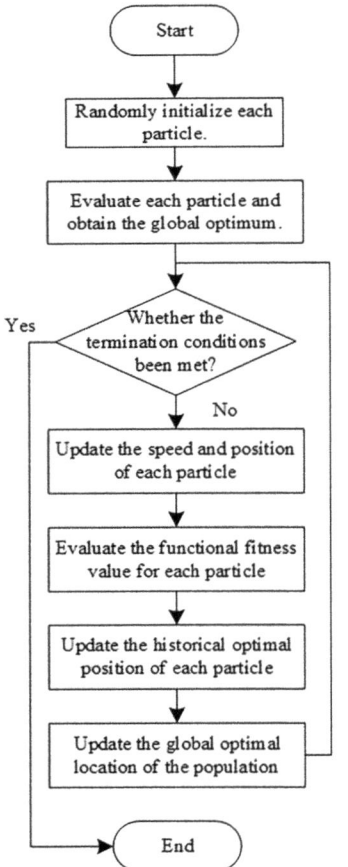

```
procedure PSO
    for each particle i
        Initialize velocity Vi and position Xi for particle I
        Evaluate particle i and set pBesti, =Xi
    end for
    gBest = min {pBesti}
    while not stop
        for i=1 to N
            Update the velocity and position of Particle i
            Evaluate particle i
            if fit(Xi) < fit(pBest,)
                pBest =X;
            if fit(pBest, ) < fit (gBest)
                gBest= pBest ;
        end for
    end while
    print gBest
end procedure
```

Fig. 3.4 Flow chart of the algorithm

(Chromosome), and the choice of fitness function is generally determined according to the research purpose. For example, in classification problems, the rate of misclassification can be used as the fitness value.

Example 3.1 The benchmark function illustrates the process of PSO algorithm to train the neural network. In the dataset, each set of data contains four attributes of the iris: sepal length, sepal width, petal length, and petal width. Three different flowers each had 50 sets of data, for a total of 150 sets of data or patterns.

Solution Using the neural network of 3 layers for classification, the network has 4 inputs and 3 outputs, so the input layer of the neural network has 4 nodes, and the output layer has 3 nodes. The number of hidden layer nodes can be dynamically adjusted. Here, it is assumed that the hidden layer has 6 nodes. All the parameters in the neural network can be trained, and only the network weights are determined here. Particles represent a set of weights of the neural network, which should be $4 \times 6 + 6 \times 3 = 42$ parameters. The range of the weights was set to $[-100, 100]$. The fitness function also needs to be determined after completing the encoding. For the classification problem, all the data are sent into the neural network, and the network weight is determined by the parameters of the particles. The number of all misclassifications was recorded as the adaptation value of the particles. The lowest possible number of mis-classifications can be obtained by training neural networks using the PSO algorithm. The PSO algorithm itself does not have many parameters to adjust, so in the experiment, adjust only the number of nodes in the hidden layer and the range of weights to achieve good classification effect.

3.1.3 Classification of the Particle Group Algorithm

Deformation of the Standard PSO Algorithm

In the deformation of the standard PSO algorithm, the standard PSO, the inertia factor, the convergence factor, the learning factor c_1 in the "cognitive" part and the "social" part are mainly changed and adjusted, hoping to obtain good results.

The original version of the inertia factor remained unchanged, and it was later proposed that the inertia factor needs to decrease gradually as the algorithm iterates on. At the beginning of the algorithm, the large inertia factor can make the algorithm not easy to fall into the local optimum. In the later stage of the algorithm, the small inertia factor can accelerate the convergence speed and make the convergence more stable, and avoid the oscillation phenomenon. After testing, dynamically reducing the inertia factor w, it can indeed make the algorithm more stable, and the effect is better. But what is the method of the decreasing inertia factor? The first thing people think of is linear decline, which is a really good strategy, but is it optimal? Therefore, some people do relevant research on the decreasing strategy, and the results point out that the decrease of linear function is better than the decreasing strategy of convex function, but the decreasing strategy of concave function is better than the decrease of linear. After testing, the experimental results basically accord with this conclusion, but the effect is not very obvious.

For the convergence factor, it has been proven that if the convergence factor is set to 0.729 can ensure the convergence of the algorithm, but it cannot guarantee that the algorithm will converge to the global optimum. After testing, a convergence factor of 0.729 has shown better results. Some people have also proposed for learning factors c_2 and c_1: The idea of c1 being big first and then small, while c_2 being small first and then big. Because in the early stages of algorithm operation, each particle needs to have a large cognitive part and a relatively small social part, which is similar to the situation of a group of people searching for things. In the early stages of searching for things, people basically rely on their own knowledge to find them. As they accumulate more and more experience, everyone gradually reaches a consensus (social knowledge), and thus begins to rely on social knowledge to find things.

In 2007, two Greek scholars proposed the method of combining the global version of PSO algorithm which is relatively fast convergence with the local version of PSO algorithm which is not easy to fall into local optimal. The speed update formula and the position update formula are respectively

$$v = n^* v_{\text{Global version}} + (1-n) * v_{\text{Local version}} \tag{3.3}$$

$$w(k+1) = w(k) + v \tag{3.4}$$

The algorithm discusses in the literature the cases where the coefficients n take various different cases and the results of analyzing various coefficients after running nearly 20,000 times.

Mix of the PSO Algorithm

The PSO algorithm can be mixed with various algorithms, such as with simulated annealing algorithms or with simplex methods. However, the most used is to combine PSO algorithm with genetic algorithm, so that three different hybrid algorithms can be generated according to three different operators of genetic algorithm.

1. the combination of PSO algorithm and the selection operator, combined with ideas: in the original PSO algorithm select population optimal value as pg, but the combined version is based on the size of all particle fitness to each particle gives a selected probability, then according to the probability of the particles, the selected particles as the optimal, other cases remain the same. Such an algorithm can maintain the diversity of the particle swarm during the operation of the algorithm, but the fatal disadvantage is the slow convergence rate.
2. The combination of the PSO algorithm and the cross operator, combined with the idea: in the process of running the algorithm, according to the size of the fitness, the particles can be crossed in two pairs, such as a very simple formula:

$$W(\text{new}) = n^* w_1 + (1-n) w_2 \tag{3.5}$$

3.1 Particle Swarm Optimization

w_1 and w_2 are the parent particles of this new particle. This algorithm can introduce new particles in the running process of the algorithm, but once the algorithm falls into the local optimum, then the particle swarm algorithm will be difficult to get rid of the local optimum.
3. The combination of PSO algorithm and mutation operator, combined with the idea: test the distance between all particles and the current optimal, when the distance is less than a certain value, some particles can be randomly initialized, so that these particles can find the optimal value again.

Binary PSO Algorithm

The original PSO algorithm was developed from solving the continuous optimization problem. Eberhart et al. also proposed the discrete binary version of PSO algorithm to solve the combinatorial optimization problem in engineering practice (Eberhart and Shi 1998). In the proposed model, they limit the historical and global optimum of each dimension of the particle to 1 or 0, without the speed. When updating the position with the speed, a threshold value is set, and when the speed is above that threshold, the position of the particle is taken as 1, otherwise 0 is taken. The binary PSO algorithm is formally similar to the genetic algorithm, but the experimental results show that the binary PSO algorithm is faster than the genetics algorithm in most test functions, especially when the dimensionality of the problem increases.

Collaborative PSO Algorithm

The cooperative PSO algorithm divides the D dimension of the particles into D populations, each optimizes the one dimensional vector, and combines these components into a complete vector when evaluating fitness. For example, for the i-th particle population, except for the i-th component, the other D-1 components are set to the optimal value, and the i-th component is constantly replaced with the i-th particle population until the optimal value of the i-th dimension is obtained, and the other dimensions are the same. In order to divide the related components into a group, the D dimension vector can be assigned to m particle group optimization, and the dimension of the first (D mod m) particle group is D/m. The dimension of the posterior m-(D mod m) swarm is the downward integer of D/m. The cooperative PSO algorithm has faster convergence rates on some problems, but the algorithm is prone to deception.

Mixed-Strategy PSO Algorithm

Hybrid strategy Hybrid PSO algorithm is to apply other evolutionary algorithms, traditional optimization algorithms or other technologies to the PSO algorithm to improve the particle diversity, enhance the global exploration ability of particles or improve the local development ability, and enhance the convergence speed and accuracy. There are usually two kinds of hybrid strategies: one is to use other optimization techniques to adaptively adjust the shrinkage factor/inertia value, acceleration constant, etc.; the other is to combine PSO algorithm with other evolutionary algorithms, such as mixing ant algorithm and PSO algorithm to solve the discrete optimization problem.

Robinson and Jiang combine the genetic algorithm and PSO algorithm for antenna optimization design and recurrent neural network design, respectively, and divide the population dynamics into multiple sub-populations, and then independently evolve the different sub-populations using PSO algorithm, genetic algorithm or mountain climbing method. Naka et al. introduced the selection operation in the genetic algorithm into the PSO algorithm to replicate the better individuals at a certain selection rate. Angeline Championship selection selection PSO algorithm, according to the fitness of the current position of the individual, each individual compared with several other individuals, and then according to the comparison of the comparison of the whole group, with the best half of the current position and speed of the worst half position and speed, while retaining the individual individual individual best position. Cross-operation of particle position and velocity by EDib et al. Higashi The Gaussian variation was introduced into the PSO algorithm. Miranda et al. use multiple operations of variation, selection and reproduction to simultaneously adaptively determine the neighborhood best position in the speed update formula, as well as the inertial weights and acceleration constants. By selecting the particle best position in the speed update formula using the differential evolution operation, Zhang et al. However, Kannan et al. optimized the inertial weight and acceleration constant of the PSO algorithm using differential evolution.

Common mixed-strategy PSO algorithms mainly include the following types.

1. The Gaussian PSO algorithm. Since the traditional PSO algorithm tends to search in the middle of the global and local optimal positions, the search ability and convergence performance depend heavily on the setting of the acceleration constant and inertia weight. In order to overcome this deficiency, Secrest et al. introduce the Gaussian function into the PSO algorithm to guide the motion of particles. The GPSO algorithm no longer requires inertial weights, and the acceleration constant is generated by random numbers following a Gaussian distribution.
2. The stretching PSO (SPSO) algorithm applies the so-called stretching technology and deflection and rejection technology to the PSO algorithm to transform the objective function and limit the movement of the particle to the local minimum solution that has been found, so that the particle has more opportunities to find the global optimal solution.
3. Chaos PSO (CPSO) algorithm. Chaos is a common nonlinear phenomenon in nature that appears chaotic but actually implies inherent regularity, characterized by randomness, ergodicity, and regularity. The CPSO algorithm is formed by generating a chaotic sequence based on the historical best position of the particle swarm, and randomly replacing the position of a particle in the particle swarm with the best position in this sequence. There is also an adaptive PSO algorithm

that uses inertia weights to adapt to the objective function value for global search, and a chaotic PSO algorithm that uses chaotic local search to locally search for the optimal position.
4. Immune-PSO algorithm. The biological immune system is a nonlinear system that is highly robust, distributed, adaptive and equipped with strong recognition ability, learning and memory ability. Existing literature has introduced the immune information processing mechanism (immune diversity, immune memory, immune self-regulation) of the immune system into the PSO algorithm, and put forward the immune PSO algorithm based on vaccination and immune memory respectively.
5. Quantum PSO algorithm optimizes to apply quantum individuals to the discrete PSO algorithm and update the particle positions based on quantum behavior.

3.1.4 Study on the Improvement of the Particle Swarm Optimization Algorithm

The research hotspot and direction of PSO algorithm include algorithm theory research, algorithm parameter research, topology structure research, hybrid algorithm research, algorithm application research, etc.

Improvement of Theoretical Research

Clerc and Kennedy designed a parameter called the compression factor in 2002. After using this parameter, the PSO algorithm converges faster (Clerc and Kennedy 2002). Trelea In 2003, it was stated that the PSO algorithm eventually converges steadily to a certain point in the space, but it is not guaranteed to be a global optimum (Trelea 2003). Kadirkamanathan et al., study the behavior of PSO algorithm in dynamic environment, from static analysis to dynamic analysis. Van den Bergh et al. tracked the flight trajectory of PSO algorithm and went deep into the dynamic system analysis and convergence study.

Improvement of the Topology Structure

Improvements in the topology of the PSO algorithm include the following.
1. Static topology structure, divided into global version (such as star structure) and local version (such as ring structure, toothed structure, von Neumann structure), etc. Figure 3.5 gives four typical topology structures.
2. Dynamic topology, including the gradual growth method proposed by Suganthan in 1999, the minimum distance method proposed by Hu and Eberhart in 2002, the recombination method proposed by Liang and Suganthan in 2005, and the random selection method proposed by Kennedy in 2006.

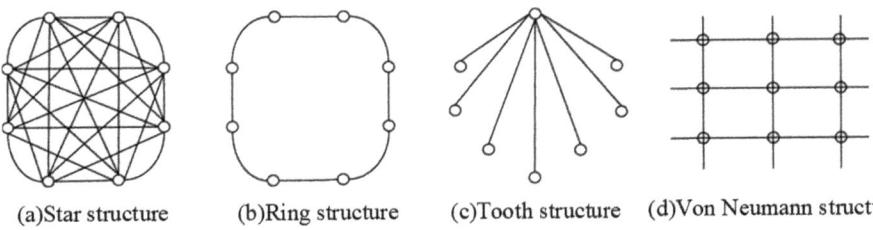

(a)Star structure (b)Ring structure (c)Tooth structure (d)Von Neumann struct

Fig. 3.5 Schematic representation of the 4 typical topology

3. Other topological structures, including the social convergence method proposed by Kennedy, Fully Informed and Liang, and the extensive learning strategies proposed by Mendes, etc.

For the currently widely used static topology structures, there are differences in convergence characteristics between the Global PSO (GPSO) algorithm and the Local PSO (LPSO) algorithm.

1. The GPSO algorithm has high connectivity and often converges faster than the LPSO algorithm. However, rapid convergence also comes at the cost of rapidly decreasing diversity for the GPSO algorithm.
2. The LPSO algorithm, due to its better diversity, is generally less likely to fall into local optima and has better performance in dealing with multimodal problems.

The following rules are followed when using static topologies to solve specific problems.

1. The topology with small neighbors has advantages in handling complex and multi-peak problems, such as the LPSO algorithm for ring structures.
2. With the expansion of the neighborhood, the convergence rate of the algorithm will be accelerated, which is very beneficial for simple, single-peak problems, such as the GPSO algorithm performs very well in these problems.

Improvement of the Hybrid Algorithm

1. Improvement of the hybrid evolutionary operator: selection operator, crossover operator, variation operator, evolutionary planning, and evolutionary strategy.
2. Improvement of mixed other search algorithms: combined with simulated annealing algorithm, artificial immune algorithm, differential evolution algorithm or local search algorithm, etc.
3. Improvement of mixed other technologies: including simplex technology, function extension technology, chaos technology, quantum technology, collaborative technology, niche technology, and speciation technology.
4. Binary coding: In 1997, Kennedy and Eberhart discretized the PSO algorithm and formed the binary coding PSO algorithm, and achieved good results in testing the five standard test functions proposed by De Jong.

5. Integer code: Salman et al. rounded the position variable of the particle to the closest legitimate discrete value. Yoshida et.al divide the continuous value fields into multiple intervals, each giving a corresponding discrete value.
6. Other forms: Schoofs and Naudts redefine the "addition, subtraction and multiplication" method of the PSO algorithm and apply it to the constraint satisfaction problem (Constraint Satisfaction Problem, CSP). Hu et al. defines speed as the probability of position variables exchanging with each other, thus discretizing the PSO algorithm to solve the n-queen (n-queen) problem. Clerc defines the appropriate "addition, subtraction and multiplication" method for the PSO algorithm, and it has applied it to solve the travel quotient problem. Chen et al. Based on the set theory technique, he redefined the speed and position update formula of the PSO algorithm and realized the discretization.

3.1.5 Parameter Settings of the Particle Swarm Optimization Algorithm

The Population Size of N

The population size affects the search capability and computational complexity of the algorithm, The PSO algorithm does not require a high population size, and generally takes 20 ~ 40 to achieve good solving results. However, for more difficult problems or specific categories of problems, the number of particles can be set to 100 or 200. The length D of the particle is determined by the optimization problem itself, which is the length of the solution to the problem. The range R of particles is determined by the optimization problem itself, and different ranges can be set for each dimension.

Maximum Speed V_{max}

The maximum speed determines the maximum distance that particles can move each time, which constrains the exploration and development capabilities of the algorithm. Each dimension V of V_{max} can generally take 10–20%, or even 100%, of the corresponding search space. There are also studies that use a setting scheme of decreasing V_{max} from large to small according to evolutionary algebra.

Inertial Weight ω

Inertial weight control before the influence of the speed on the current speed, to balance the algorithm exploration and development ability generally set from 0.9 to 0.4, also have nonlinear decreasing setting scheme can be set by fuzzy control, or randomly between [0.5, 1.0], ω set to 0.729 while c_1 and c_2 to 1.49445, conducive to the convergence of the algorithm.

Compression Factor χ

The compression factor can limit the flight speed of the particles and ensure the effective convergence of the algorithm. Clerc et al. mathematically obtained the χ value of 0.729 with c_1 and c_2 set to 2.05.

Acceleration Coefficients c1 and c2

The acceleration coefficients c1 and c1 represent the acceleration weights of the particle advancing to its own extreme pBest and global extreme gBest, c_1 is the individual learning factor of the particle and c_2 is the social learning factor of the particle. Both c_1 and c_2 are each equal to 2.0, representing equal emphasis on both guidance directions, and there are also some unequal settings for c_1 and c_2, but they generally range from 0 to 4. Studying the adaptive adjustment scheme of c_1 and c_2 is important for the enhancement of the algorithm performance.

Termination Conditions

Termination conditions determine the end of the algorithm running, determined by the specific application and the problem itself will set the maximum number of cycles to 500, 1000, 5000 or the maximum number of function evaluation, also can use the algorithm to get an acceptable solution as a termination condition or when the algorithm in a long iteration without any improvement, can terminate the algorithm.

Global and Local PSO

Global and local PSO decide how to choose the two versions of PSO—GPSO and LPSO, GPSO is fast, but sometimes falls into local optimum; LPSO convergence is a little slower, but it is not easy to fall into local optimum. In practice, the specific algorithm version can be selected according to the specific problem.

Synchronous and Asynchronous Updates

The difference between the two update methods is the update method for global gBest or local pBest. In the synchronous update method, in each generation, when all particles use the current gBest for speed and position updates, the particles are evaluated, their respective pBest is updated, and the best pBest is selected as the new gBest.

3.1 Particle Swarm Optimization

In asynchronous update mode, in each generation, particles use the current gBest for speed and position updates, and then immediately evaluate and update their own pBest. If their pBest is better than the current gBest, gBest is immediately updated to quickly use the better gBest for the subsequent particle update process.

Generally speaking, PSO algorithms with asynchronous updates have efficient information propagation capabilities and faster convergence speeds.

3.1.6 Comparison of the Particle Swarm Optimization Algorithm and the Genetic Algorithm

The process of the PSO algorithm and the genetic algorithm generally follows the following steps.

1. Population random initialization.
2. Calculate the fitness values for each individual within the population. The fitness value is directly related to the distance of the optimal solution.
3. Population are replicated according to fitness values.
4. If the termination conditions are met, stop, otherwise go to step 2.

From the above steps, we can see that the PSO algorithms and the genetic algorithms have a lot in common. Both can randomly initialize the population randomly, and both evaluate the system using fitness values, and both search randomly against the fitness values. Neither system guarantees that the optimal solution must be found. However, instead of genetic operations such as crossover and variation, the PSO algorithm determines the search according to its own speed. At the same time, the particle of the PSO algorithm also has an important feature, which is the memory.

The mechanism of information sharing in PSO is different than that of the genetic algorithm. In the genetic algorithm, chromosomes share information with each other, so the whole population moves more evenly to the optimal region. In the PSO algorithm, only gBest information is given to other particles, which is a one-way information flow, and the whole search update process is a process that follows the current optimal solution. Compared with GA, in most cases, all particles may converge to the optimal solution faster.

The advantage of evolutionary computing is that it can deal with some problems that cannot be handled by traditional methods, such as non-inducible node transfer functions or no gradient information. However, it may also occur that the performance is not particularly good, and the selection of genetic operators is also troublesome.

Currently, there are some papers on training neural networks using PSO algorithm instead of back-propagation algorithm. Research show that PSO is a promising neural network algorithm.

3.1.7 Related Applications of Particle Swarm Optimization Algorithm and MATLAB Examples

The PSO Algorithm for Its Application

The PSO algorithm uses birds to simulate particles. Each particle can be regarded as a search individual in the N-dimensional search space. The current position of the particle is a candidate solution to the corresponding optimization problem, and the flight process of the particle is the search process of the individual. The flight speed of particles can be dynamically adjusted according to the optimal particle history and the optimal population history. The particle has only two properties: speed and position, speed represents the speed of movement, and position represents the direction of movement. The optimal solution for each particle is called individual extreme value, and the optimal individual extreme value in the particle population is taken as the current global optimal solution. The algorithm constantly iterates, updates the speed and position, and finally obtains the optimal solution that satisfies the termination condition.

The iterative process of PSO algorithm is shown in Fig. 3.6. The specific process is as follows.

1. Initialization. Firstly, set the maximum number of iterations, the number of independent variables in the objective function, the maximum degree of particles, and the position information as the entire search space. Randomly initialize the velocity and position in the velocity interval and search space. Set the particle swarm size to M and randomly initialize a flying velocity for each particle.
2. Individual extreme values and the global optimal solution. The fitness function is defined, and the individual extreme value is the optimal solution found by each particle. Find a global value from these optimal solutions, which is called the global optimal solution, which is compared with the historical global optimum and updated.
3. The update speed and location are

$$V_d^i = \omega c_1 \text{random}(0,1)\left(\text{pBest}_i^d - X_i^d\right) + c_2 \text{random}(0,1)\left(\text{pBest}^d - X_i^d\right) \quad (3.6)$$

Among them, ω is the inertia weight, its value is non-negative, large global optimization ability is strong and local optimization ability is weak, small global optimization ability is weak and local optimization ability is strong. By sizing ω, we can adjust the global and local optimal performance. And c_1 and c_2 are the acceleration coefficient. The experiment of Suganthan show that c_1 and c_2 can be better when they are constants solution, usually set $c_1 = c_2 = 2$, but not necessarily equal to 2, generally $c_1 = c_2$ [0,4]. Random (0,1) represents a random number on the interval [0, 1], pBest_i^d represents the d-th dimension of the individual pole of the i-th variable, pBest^d represents the globally optimal solution of the d dimension.

3.1 Particle Swarm Optimization

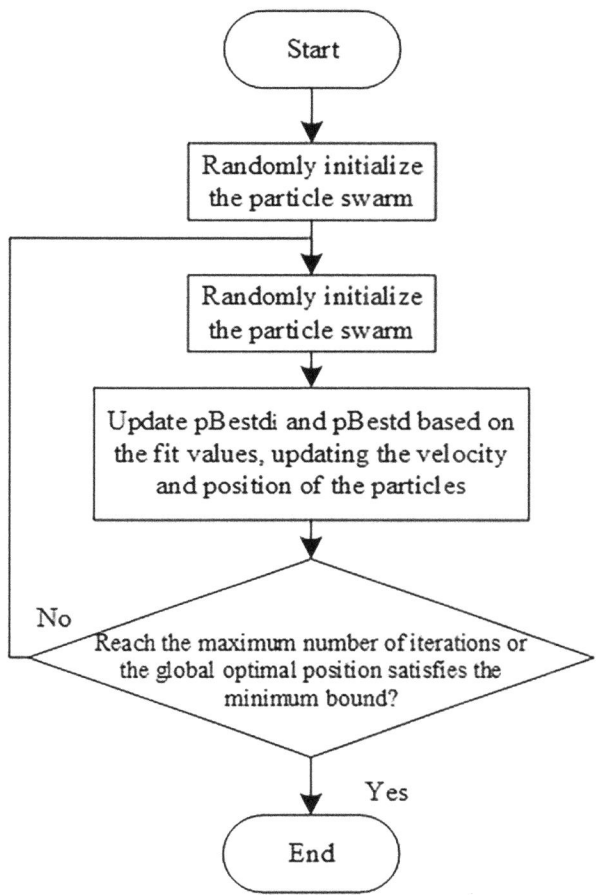

Fig. 3.6 Iterative process of PSO algorithm

4. Termination condition: reaches the set number of iterations; the difference between algebra meets the minimum limit. Figure 3.6 depicts the iterative flow of the PSO algorithm.

Some advantages of the PSO algorithm are shown as follows.

1. Is a class of uncertain algorithm. Uncertainty embodies the biological mechanism of organisms in nature and outperforms deterministic algorithms in solving certain specific problems.
2. It is a class of probabilistic global optimization algorithm. The advantage of the non-definite algorithm is that the algorithm has more opportunities to solve the global optimal solution.
3. It does not rely on the strict mathematical properties of the optimization itself.

4. It is a biomimetic optimization algorithm based on multiple agents. The various agents in the PSO algorithm can better adapt to the environment through mutual collaboration and show the ability to interact with the environment.
5. With essential parallelism. Including intrinsic parallelism and embedded parallelism.
6. Outstanding in nature. The completion of the total goal of the PSO algorithm emerges during the movement of the individual behavior of multiple agents.
7. With self-organization and evolvability and memory function, all the particles preserve the relevant knowledge of the optimal solution.
8. Robustness. Robustness refers to the practicability and effectiveness of the algorithm under different conditions and environments. However, the mathematical theoretical foundation of PSO algorithm is not solid enough, and the convergence of the algorithm needs to be discussed.

The MATLAB Example of the PSO Algorithm

Example 3.2 Find the optimal solution for function $f(x) = x\sin x\cos 2x - x\sin 3x$

First, initialize the population. Given the position constraints of [0, 20], due to the simplicity of one-dimensional problems, the initial population N can be set to 50, the number of iterations to 100, and the spatial dimension d = 1. The initialization of position and velocity is to randomly generate an N × d matrix within the position and velocity limits. For this function, position initialization is to randomly generate a 50 × 1 data matrix within 0–20. The position constraint here can also be understood as position constraint, while velocity constraint is to ensure that the particle step size does not exceed the limit. Generally, the velocity constraint is set to [−1, 1]. Another characteristic of particle swarm optimization is to record the historical best of each individual and the historical best of the population, so the corresponding optimal positions and values of the two also need to be initialized. The historical optimal position of each individual can be initialized as the current position, while the historical optimal position of the population can be initialized as the origin. For the optimal value, if the maximum value is sought, it is initialized to negative infinity, otherwise it is initialized to positive infinity. Each search requires comparing the current fitness value and the optimal solution with historical record values. If the historical optimal value is exceeded, the historical optimal positions of individuals and populations are updated.

Speed and position update is the core of the particle swarm algorithm. Its principle expression and update mode are as follows:

$$\begin{cases} V_i^d = w \cdot V_i^d + c_1 r_1 \left(\text{pBest}_i^d - x_i^d \right) + c_2 r_2 \left(\text{pBest}^d - x_i^d \right) \\ x_i^d = x_i^d + v_i^d \end{cases} \quad (3.7)$$

After each update of the speed and position, it is necessary to consider the speed and position restrictions, and you need to limit them to the specified range. Here, only a

3.1 Particle Swarm Optimization

conventional method is used to restrict the ultra-constrained data to the boundary. Figures 3.7, 3.8, and 3.9 show the initial state, final state and convergence process diagram of the particle, respectively. So the algorithm has successfully found the optimal solution, and the optimal solution is 18.3014, while its maximum value is 32.1462.

Neuronal Approach Combined with the PSO Algorithm

In general, the PSO algorithm can be used to conduct a global search, obtain an initial solution, and then a more careful search using the back propagation algorithm.

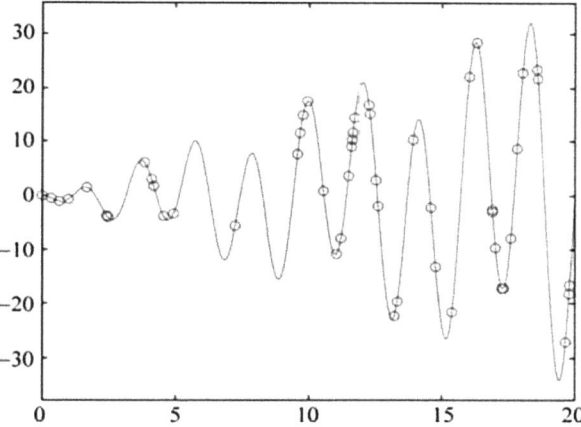

Fig. 3.7 Initial Particle state diagram

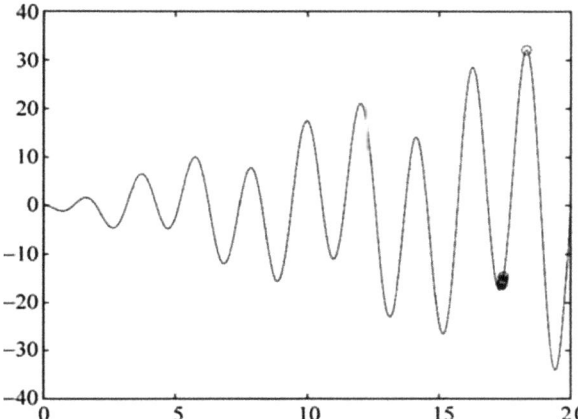

Fig. 3.8 Particle final state diagram

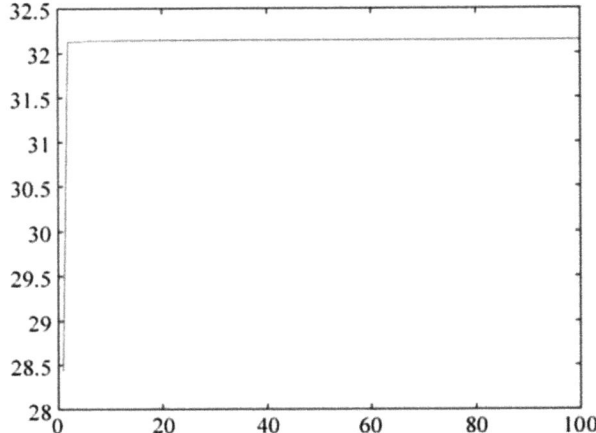

Fig. 3.9 Particle convergence process

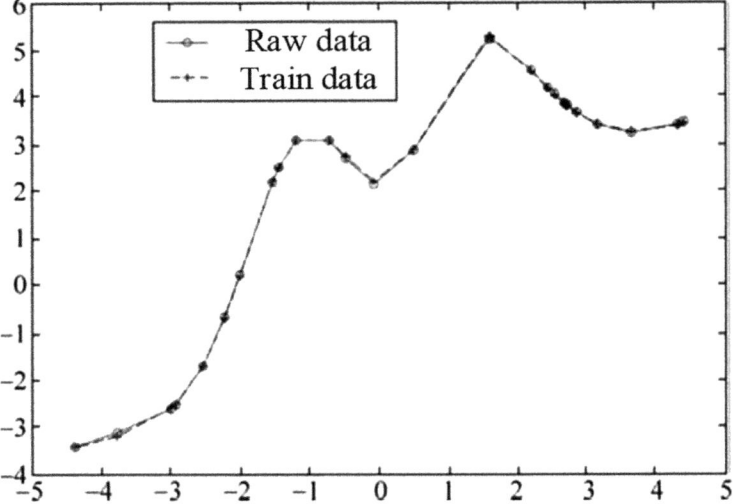

Fig. 3.10 Particle training process

Example 3.3 Sampling of the function $2.1*(1 - x + 2x)e(-1/2) + \sin x + x$, $x \in [-5, 5]$, obtain 30 sets of training data and fit the neural network.

Solution The neural network structure is 1-7-1 structure, including 1 input neuron, 7 inter-neurons and 1 output neuron. Figure 3.10 shows the training process of particles. The first step of fitting extracts 30 sets of data, including input and output; the second step runs the PSO algorithm for random search and selects an optimal solution with 22 dimensions; and the third step uses the back propagation algorithm on the basis of the PSO algorithm solution.

3.2 Ant Colony Algorithm

3.2.1 Basic Principles of the Ant Colony Algorithm

Ant Colony Algorithm (ACA) is a novel simulation evolutionary algorithm proposed in recent years. Italian scholar Dorigo first proposed ACA based on his research on the collective behavior of real ant colonies in nature (Dorigo et al. 2000).

Swarm intelligence represented by ACA has become a hot topic in distributed artificial intelligence research, and many algorithms designed from bee colonies and ant colony models are increasingly being applied to the study of enterprise operation modes. The current research on ACA not only includes algorithmic studies, but also studies from the perspective of simulation models, and scholars continue to propose improvement schemes for ant colony algorithms. From the current available literature, scholars studying and applying ant colony algorithms are mainly concentrated in European countries such as Belgium, Italy, the United Kingdom, France, and Germany. Japan and the United States have also initiated research on ant colony algorithms, and there have been a few public reports and research results in China since the end of 1998.

Swarm intelligence is also applied to the formulation of factory production plans and logistics management in transportation departments. Pacific Southwest Airlines has adopted a transportation management software directly derived from ant behavior research, saving at least $ten million in annual expenses. Unilever UK has taken the lead in using swarm intelligence technology to improve the operation of one of its toothpaste factories. General Motors, French Hydro, the Dutch Ministry of Transport, and some immigration agencies in the United States have also adopted this technology to improve their operational capabilities. British Telecom and WorldCom conducted experiments on new telecommunications network management methods based on electronic ants.

ACA was originally used to solve TSP problems, and after years of development, it has gradually extended to other fields such as graph coloring problems, large-scale integrated circuit design, routing problems in communication networks, load balancing problems, and vehicle scheduling problems. Ant colony algorithm has been successfully applied in several fields, among which the most successful application is combinatorial optimization problem.

As shown in Figs. 3.11 and 3.12, ant colonies in nature communicate indirectly with each other through a substance called pheromone while searching for food, enabling them to collaborate in discovering the shortest path from the ant nest to the food source. By abstractly modeling this type of swarm intelligence behavior, researchers have proposed the Ant Colony Optimization algorithm, which provides a powerful means for solving optimization problems, especially combinatorial optimization problems.

Ants often randomly choose their path when searching for food, but they can sense the current concentration of pheromones on the ground and tend to move towards directions with higher concentrations of pheromones. Pheromones are

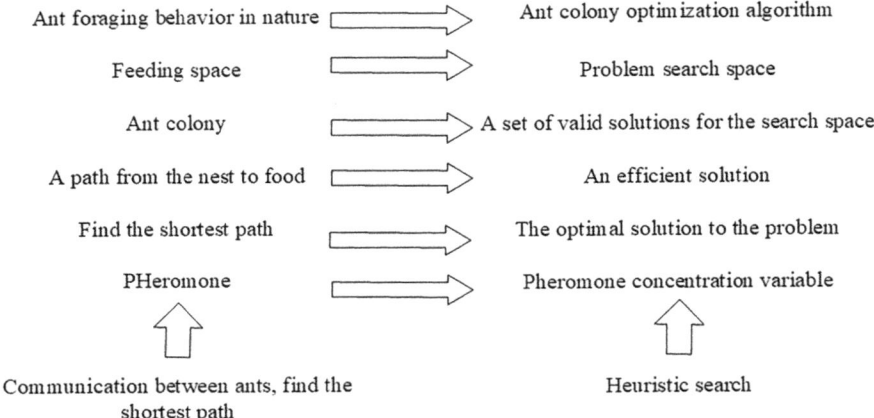

Fig. 3.11 Flow chart of the ant colony information search

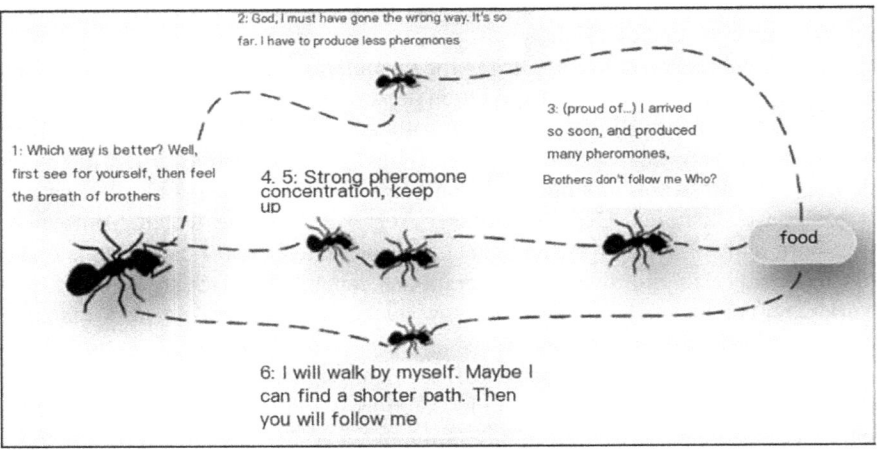

Fig. 3.12 Ant colony information search diagram

released by ants themselves and are substances that enable indirect communication within ant colonies. Due to the shorter round-trip time of ants on shorter paths and the fact that more ants pass through the path in a unit of time, the accumulation speed of pheromones is faster than on longer paths. Therefore, when the subsequent ants are at the intersection, they can perceive the information left by the previous ants and tend to choose a shorter path to move forward. This positive feedback mechanism allows more and more ants to travel along the shortest path between their nest and food. Due to the evaporation of pheromones on other paths over time, all ants ultimately travel on the optimal path.

3.2.2 Algorithm Process of the Ant Colony Algorithm

Ant system (AS) is the most basic ACO algorithm, which is proposed with TSP as an application example. The ACO algorithm contains two essential elements.

1. Path construction: Each ant randomly selects a city as its starting city, and maintains a path memory vector to store the city through which the ant passes in turn. The ants select the next city to arrive according to a random proportional rule at each step of the construction path.
2. Phereromone update: When all the ants have built their paths, the algorithm will update the global pheromone of all the paths. Note that the described is the ant-cycle version of AS, the update was performed after all the ants have completed the path construction, and the pheromone concentration change is correlated to the path length that the ants build in this round.

Path Building

For each ant k, the path memory vector $\mathbf{R}^k$ records the city sequence number that all ant k have passed through in the order of visits. Given the current city i, the probability of selecting city j as the next visiting object is

$$p_k(i,j) = \begin{cases} \dfrac{[\tau(i,j)]^\alpha [\eta(i,j)]^\beta}{\sum_{u \in J_k(i)} [\tau(i,u)]^\alpha [\eta(i,u)]^\beta}, & j \in J_k(i) \\ 0, & \text{Other} \end{cases} \quad (3.8)$$

Where $J_k(i)$ represents the collection of cities in the city sequence $\mathbf{R}^k$ that can be reached directly from the city i and is not visited by the ants. $\eta(i, j)$ is a heuristic information that is usually calculated directly by $\eta(i, j) = 1/d_{ij}$. $\tau(i, j)$ represents the amount of pheromone on the edge (i, j). α, β is two constants, the weighted values of pheromone and visibility, respectively.

Pheromone Updates

1. When the algorithm initializes, the pheromones on all the edges of the problem space are initialized as t_0.
2. For each round of algorithm iteration, the pheromones on all paths in the problem space will evaporate. We multiply a constant less than 1 for the pheromones on all edges. Pheromone evaporation is an inherent feature of nature itself, which can help avoid the infinite accumulation of pheromones in the algorithm, allowing the algorithm to quickly discard the previously constructed poor paths.

3. Ants release pheromones on the edge of their current round according to the path length they build. The shorter the path the ant builds, the more pheromones it releases. The more an edge is crawled by an ant, the more pheromones it gets.
4. Iteration step 2, until the termination of the algorithm.

$$\tau(i,j) = (1-\rho) \cdot \tau(i,j) + \sum_{k=1}^{m} \Delta\tau_k(i,j),$$

$$\Delta\tau_k(i,j) = \begin{cases} (C_k)^{-1}, & (i,j) \in R^k \\ 0, & \text{Other} \end{cases} \quad (3.9)$$

Where m is the number of ants; ρ is the evaporation rate of pheromones, specified $0 < \rho \leq 1$. "$\Delta\tau_k(i,j)$ is the amount of pheromone released by the k-th ant on the edge where it passes, which is equal to the reciprocal of the construction path length of the ant k. C k represents the path length, which is the length sum of all edges in R k.

Example of the Algorithm Process

Example 3.4 Given the execution steps to solve the TSP problem in the four cities with the ant colony algorithm, and the distance matrix between A, B, C and D of the four cities is

$$W = d_{ij} = \begin{bmatrix} \infty & 3 & 1 & 2 \\ 3 & \infty & 5 & 4 \\ 1 & 5 & \infty & 2 \\ 2 & 4 & 2 & \infty \end{bmatrix} \quad (3.10)$$

Suppose the size of the ant population m = 3, α = 1, β = 2, and ρ = 0. 5. separate:

Step 1: Initialization. Firstly, let the greedy algorithm obtain the path (ACDBA), then C^{mn} = f(ACDBA) = 1 + 2 + 4 + 3 = 10. Obtain $\tau_0 = \dfrac{m}{C^{mn}} = 0.3$. Initialize all pheromones on the edges with $\tau_{ij} = \tau_0$.

Step 2.1: Randomly select the departure city for each ant, assuming Ant 1 chooses City A, Ant 2 chooses City B, and Ant 3 chooses City D.

Step 2.2: Select the next city for each ant. Taking Ant 1 as an example, the current city is A, and the accessible city set $J_1(i)$ = {B, C, D}.Calculate the probability of Ant 1 selecting B, C, and D as the next visited city

$$A \Rightarrow \begin{cases} B: \tau_{AB}^\alpha \times \eta_{AB}^\beta = 0.3^1 \times (1/3)^2 = 0.033 \\ C: \tau_{AC}^\alpha \times \eta_{AC}^\beta = 0.3^1 \times (1/1)^2 = 0.3 \\ D: \tau_{AD}^\alpha \times \eta_{AD}^\beta = 0.3^1 \times (1/2)^2 = 0.075 \end{cases}$$

3.2 Ant Colony Algorithm

$$p(B) = 0.033 / (0.033 + 0.3 + 0.075) = 0.081$$

$$p(C) = 0.3 / (0.033 + 0.3 + 0.075) = 0.74$$

$$p(D) = 0.075 / (0.033 + 0.3 + 0.075) = 0.18$$

Use the roulette selection rule to choose the next city. Assuming the resulting random number q = random (0,1) = 0.05, then ant 1 will choose city B.
The same method was used for selecting the next visiting city for ants 2 and 3, assuming that ant 2 selects city D and ant 3 selects city A.

Step 2.3: The city where ant 1 is B, the path memory vector R1 = (AB), and the city collection J1 (i) = {C,D}. The probability of ant 1 choosing C and D as the next city was calculated as

$$B \Rightarrow \begin{cases} C : \tau_{BC}^{\alpha} \times \eta_{BC}^{\beta} = 0.3^1 \times (1/5)^2 = 0.012 \\ D : \tau_{BD}^{\alpha} \times \eta_{BD}^{\beta} = 0.3^1 \times (1/4)^2 = 0.019 \end{cases}$$

$$p(C) = 0.012 / (0.012 + 0.019) = 0.39$$

$$p(D) = 0.019 / (0.012 + 0.019) = 0.61$$

Use the roulette selection rule to choose the next city. Assuming the resulting random number q = random (0,1) = 0.67, then ant 1 will choose city D. The same method was used for selecting the next visiting city for ants 2 and 3, assuming that ant 2 selects city C and ant 3 selects city C.

Step 2.4: In fact, the path has been constructed, and the path constructed by ant 1 is (ABDCA). The path of the ant 2 construction is (BDCAB). The path constructed by ant 3 is (DACBD).

Step 3: Pheromone update. The path length constructed by each ant is $C_1 = 3 + 4 + 2 + 1 = 10$, $C_2 = 4 + 2 + 1 + 3 = 10$, $C_3 = 2 + 1 + 5 + 4 = 12$
Update the pheromone on each side:

$$\tau_{AB} = (1-\rho) \times \tau_{AB} + \sum_{k=1}^{3} \Delta \tau_{AB}^k = 0.5 \times 0.3 + (1/10 + 1/10) = 0.35$$

$$\tau_{AC} = (1-\rho) \times \tau_{AC} + \sum_{k=1}^{3} \Delta \tau_{AC}^k = 0.5 \times 0.3 + (1/12) = 0.16$$

······

The updated amount of pheromone of all edges in the problem space is calculated according to Eq. (3.9).

Step 4: If the end condition is met, output the global optimal result and end the program. Otherwise, continue to turn to step 2.1.

3.2.3 Development of the Ant Colony Algorithm

Parameter Setting of the Ant Colony Algorithm

The parameter settings of ant colony algorithm have a significant impact on its performance. Solion analyzed the impact of pheromone related parameters on "exploration" and "mining", and proposed adding a pre-processing stage before running the ant colony algorithm. In this stage, a certain number of paths (i.e. loops) are found without using pheromones, and then some paths are selected to initialize pheromones before the algorithm starts, achieving good results. Different parameter combinations affect the performance of ant colony algorithm. Pia and Gaertner combined genetic algorithm with ant colony algorithm and applied genetic algorithm to optimize the parameters of ant colony algorithm, achieving good results. Meyerl studied the influence of parameters in ant colony algorithm on the "exploration" behavior, that is, maintaining the diversity of solutions. He pointed out that β not only plays a decisive role in coordinating the "exploration" and "mining" behaviors, but also has an important impact on the robustness of the system. There is relatively little research on the optimization of ant colony algorithm parameters. However, the parameter setting of ant colony algorithm is related to the final performance of the algorithm, so the principle of parameter setting deserves further study.

The influence of number on "exploration" and "exploitation" suggests that a pre-processing stage is added before the ant colony algorithm runs, which starts without using pheromones to find a certain number of paths (i.e. circuits), and then selects some paths to initialize the pheromone before the algorithm starts, which achieves good results. Different parameter combinations affect the performance of ant colony algorithm. Pia and Gaertner combined the genetic algorithm with ant colony algorithm and applied the genetic algorithm to optimize the parameters of ant colony algorithm, and obtained good results. Meyerl Study the parameters in the colony algorithm of "exploration" behavior, namely keep the influence of solution diversity, pointed out that β not only for the coordination of "exploration" and "mining" behavior, but also has an important impact on the robustness of the system about the ant algorithm parameter optimization research is relatively few, but the parameter setting of colony algorithm to the performance of the algorithm, so the parameter setting principle is worth more study.

Improvement of the Ant Colony Algorithm

In order to improve the performance of ant colony algorithm and obtain faster convergence rate and solution quality, many research work revolves around the improvement of ant colony algorithm. Mainly includes the following improvements, Gambardella will ants system and enhance learning Q-learning algorithm together AntQ algorithm, Dorigo with elite strategy (elitist strategy) to ant system pheromone update mechanism, which enhance the importance of find the optimal path of ants in each iteration, the path to find additional pheromone, this strategy improves the ant system ability to solve large-scale problems. Taillard et al. proposed the concept of fast ant system (Fast Ant System, FAS). FAS introduced the pheromone reset mechanism to avoid the algorithm convergence to local optima.

Stutzle Has proposed the concept of maximum-minimum ant system (MMAS), which is similar to the ant colony system introduced in Sect. 3.2.2, but the change range of pheromones on MMAS prior defined path is [τ min, τ max], and τ min and τ max are preset parameters, which can effectively avoid search stagnation. The original ant colony algorithm is suitable to solve the combinatorial optimization problem, and now some scholars try to improve the ant colony algorithm to solve the continuous optimization problem. Pourtakdoust et.al proposed an ant colony algorithm for solving continuous optimization problems only guided by pheromones. Dec et al. proposed the continuous interactive ant colony (CIAC) algorithm to solve the multi-objective continuous function optimization problem. There are relatively few studies to solve continuous optimization problems, which is a potential research direction.

Proof of the Convergence of the Ant Colony Algorithm

T. Stutztle et al. have demonstrated the algorithm convergence of MMAS, and W. Gutjahr has proved that the ants of GBAS (Graph-Based Ant System) can converge to a given optimal solution with arbitrary proximity probability. However, the current convergence proof of MMAS does not give an estimate of the convergence rate, while the execution of GBAS has more limitations than the ant colony, and has not been applied in practical combinatorial optimization problems.

Fusion of the Ant Colony Algorithm and Other Algorithms

Ant colony algorithm is easy to integrate with other algorithms to learn from each other and improve its performance. The present research results in this field include the fusion research between ant colony algorithm and genetic algorithm, neural network, micro-particle group algorithm, etc. Ding Jian et al. used the genetic method to generate the initial pheromone distribution of pathways and obtained good fruit. For the integration of ant colony algorithm and genetic algorithm, the research mainly uses genetic algorithm to optimize the parameters and pheromones in ant colony algorithm, and integrates the selection, crossover and variation in genetic algorithm into ant colony algorithm. The fusion of ant colony algorithm and neural network, Blum et al Neural network, and apply it to pattern classification.

The ant colony algorithm can also be integrated with optimization algorithms such as immunization algorithm and PSO algorithm. Holden et al., integrated the ant colony algorithm and the PSO algorithm to handle the hierarchical classification of biological datasets. Blo et al. proposed a model based on the ant colony algorithm and the rough set method, and used it for feature selection. As a new biomimetic optimization method, the ant colony algorithm is still in the infancy of comparing with other algorithms, and there are few literature and reports. Therefore, it is very meaningful to design new fusion strategies combined with other algorithms to further improve the performance of the ant colony method and improve the connection between the ant colony algorithm and other algorithms.

3.2.4 Improvement of the Ant Colony Algorithm

Essence Ant System

Elite Ant System (EAS) is the first improvement of the basic AS, which adds a means to strengthen the current optimal path on the basis of the original AS pheromone update principle:

$$\tau(i,j) = (1-\rho)\cdot\tau(i,j) + \sum_{k=1}^{m}\Delta\tau_k(i,j) + e\Delta\tau_b(i,j),$$

$$\Delta\tau_k(i,j) = \begin{cases} C_k, & if\ (i,j) \in R^k \\ 0, & otherwise \end{cases}$$

$$\Delta\tau_b(i,j) = \begin{cases} C_b, & if\ (i,j)\ \text{on path}\ T_b \\ 0, & otherwise \end{cases}$$

The introduction of this additional pheromone reinforcement means helps to better guide the bias of the ant search, allowing the algorithm to converge faster.

Permutation-Based Ant System

The arrangement-based ant system (AS_{rank}) adds a weight to the pheromones to be released on the basis of AS, further increasing the difference in the amount of pheromones on each side to guide the search. After each round of all ants building paths, they are ranked according to the length of the resulting path, and only the ants with the current optimal path and the ants ranked in the top ($\omega - 1$) are allowed to release pheromones. The weight of pheromones released by ants on edges (i, j) is determined by the ranking of the ants.

$$\tau(i,j) = (1-\rho)\cdot\tau(i,j) + \sum_{k=1}^{\omega-1}(\omega-k)\Delta\tau_k(i,j) + \omega\Delta\tau_b(i,j),$$

$$\Delta\tau_k(i,j) = \begin{cases} C_k, & (i,j) \in R^k \\ 0, & otherwise \end{cases}$$

$$\Delta\tau_b(i,j) = \begin{cases} C_b, & (i,j)\ \text{on path}\ T_b \\ 0, & otherwise \end{cases}$$

3.2 Ant Colony Algorithm

Weights ($\omega - k$) amplify the difference of pheromone concentration between the different pathways, and AS_{rank} can guide the ant search more vigorously.

Max. and Minimum Ant System

MMAS made the following improvements based on the basic AS algorithm.

1. Only allow the iterative optimal ant (the ant with the shortest path built in this iteration), or only the optimal ant to release pheromones.
2. The value range of the pheromone quantity is limited to an interval.
3. The initial pheromone value is the upper limit of the pheromone value interval, accompanied by a small pheromone evaporation rate.
4. Whenever the system enters a stagnant state, the amount of pheromone on all sides of the problem space is reinitialized.

Ant Group System

In 1997, Dorigo, the founder of the ant colony algorithm, proposed an ant colony system with a new mechanism in the article "*Ant colony system: a cooperative learning approach to the traveling salesman problem*", which further improved the performance of the ACO algorithm.

ACS is a landmark work in the history of ant colony algorithm.

1. A pseudo-random proportion rule is used to select the city nodes to establish a balance between the current development path and exploring the new path.

$$j = \begin{cases} \arg\max_{j \in J_k(i)} \left\{ [\tau(i,j)], [\eta(i,j)]^\beta \right\}, & \text{if } q \leq q_0 \\ S, & \text{otherwise} \end{cases}$$

The q0 is a parameter in the [0,1] interval. When the resulting random number q q0, the ant directly chooses the lower city node with the largest exponential product of heuristic information and pheromone quantity, which is usually called development (exploitation). On the other hand, when the resulting random number $q > q_0$, ACS will use the roulette selection strategy like various AS algorithms, which we call biased exploration.

By adjusting q_0, we can effectively adjust the balance between "development" and "exploration" to determine whether the algorithm focuses on developing the region near the optimal path or exploring other regions.

  (a) Using the pheromone global update rule, all ants in each iteration have built the path, and evaporated and released the pheromone on the edge of the hitherto optimal path.

$$\tau(i,j) = (1-\rho) \cdot \tau(i,j) + \rho \cdot \Delta\tau_b(i,j), \quad \forall (i,j) \in T_b$$

$$\Delta\tau_b(i,j) = 1/C_b$$

Both the evaporation and release of pheromones are only carried out on the edge of the optimal path to date, which is very different from AS. Because the AS algorithm applies pheromone updates to all sides of the system, the computational complexity of pheromone updates is $O(n^2)$, while the pheromone update calculation complexity of the ACS algorithm is reduced to $O(n)$. The parameter r represents the rate of pheromone evaporation. After the newly added pheromone is multiplied by the coefficient r, the updated pheromone concentration is controlled between the old pheromone amount and the newly released pheromone amount, and the limit of the pheromone amount in MMAS algorithm is realized in an implicit and simpler way.

(b) Introduce the pheromone local update rule. In the process of path construction, whenever every ant passes through an edge (i, j), it will immediately update the pheromone to this edge.

$$\tau(i, j) = (1-\xi) \cdot \tau(i, j) + \xi \cdot \tau_0 \tag{3.11}$$

The action of the pheromone local update rule on an edge reduces the probability that this edge is selected by other ants. This mechanism greatly increases the exploration ability of the algorithm, and subsequent ants tend to explore unused edges, effectively avoiding the algorithm from entering a stagnant state (Fig. 3.13).

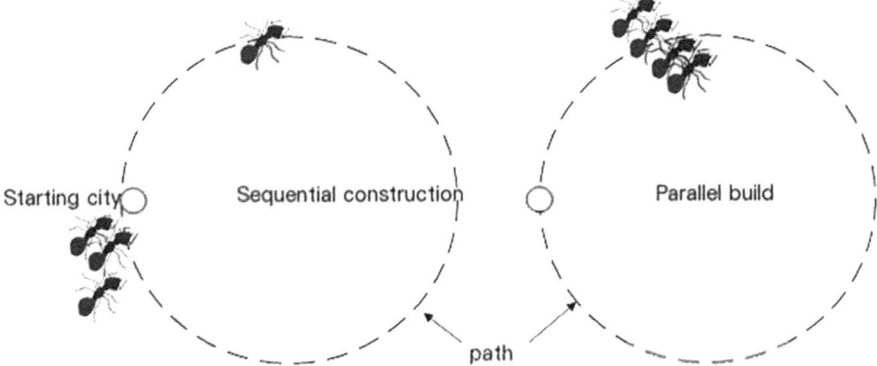

Fig. 3.13 Ant colony search Fig

3.2 Ant Colony Algorithm

Continuous Orthogonal Ant Colony System

In recent years, the ant colony algorithms that extend the application field to the continuous space are also developing, and the continuous orthogonal ant colony (COAC) algorithm is one of the more excellent The COAC algorithm searches for the optimal solution in the problem space by adaptively selecting and adjusting a certain number of regions in the problem space, and using ants conducting an orthogonal search within these regions, transferring states between regions, and updating the pheromones of each region. The basic idea of COAC is to discretize the continuous space using the method of orthogonal tests.

3.2.5 *Parameter Setting of the Ant Colony Algorithm*

The parameter settings of the ant colony algorithm are shown in Table 3.1.

3.2.6 *Application of the Ant Colony Algorithm*

Since the success of ACO algorithm in the combined optimization of some classical combined programming problems such as TSP and QAP, it has been applied to many new practical engineering fields.

Table 3.1 Parameter setting table of the ant colony algorithm

Parameter	Reference settings
Number of ants: m	With AS algorithm, EAS algorithm, AS_{rank} algorithm and MMAS algorithm to solve the TSP problem, m is equal to the number of cities n, the algorithm has good performance; but for ACS algorithm, $m = 10$ is appropriate
The pheromone weight α and the heuristic information weight β	It is more appropriate to set $\alpha = 1$ and $\beta = 2 \sim 5$ in various ACO algorithms
The pheromone volatile factor r	For AS and EAS, $r = 0.5$; $r = 0.1$ for AS_{rank}; $r = 0.02$ for MMAS and $r = 0.1$ for ACS, with high comprehensive performance for the algorithm
The initial pheromone amount t_0	For AS algorithm, $t_0 = m/C^{nn}$; for EAS algorithm, $t_0 = (e + m)/rC^{nn}$; for AS_{rank} algorithm, $t_0 = 0.5r(r-1)/rC^{nn}$; $t_0 = 1/rC^{nn}$ for MMAS; $t_0 = 1/nC^{nn}$ for ACS
Number of ants releasing the pheromones ω	In the AS_{rank} algorithm, the parameter ω is set to $\omega = 6$
Evolutionary stasis determines the algebra r_s	In the MMAS algorithm, the parameter r_s was set to $r_s = 25$
The pheromone local volatile factor x	In the ACS algorithm, the parameter x was set to $x = 0.1$
The pseudo-random factor, q_0	In the ACS algorithm, the parameter q_0 is set to $q_0 = 0.1$

1. Applications in various engineering and industrial production, such as using the idea of ACO algorithm to solve the integrated wiring problem of large-scale integrated circuit. During the wiring process, the attraction of each pin to the ant can be calculated according to the gravitational function. According to the enlightening strategy, each line network agent craw on the switch box grid like an ant colony, with a metal wire where it passes. After all the pins of a line network, the line network is distributed.
2. The application of ACO algorithm in various practical planning problems, such as the application of robot path planning. As an agent, the problem of path planning in complex working environment and the problem of cooperative strategy between multiple robots are largely similar to the preferred path of ant foraging and the cooperation of pheromones between individuals in the ant group. The path planning algorithm is one of the bases of realizing robot control and navigation, which has great advantages to use ACO algorithm to solve this problem.

In addition, ACO algorithm is also applied in dynamic optimization combination problems, specifically in directed connected network routing and unconnected network system routing. Other applications include ant artificial neural network, Vehicle Routine Problem (VRP), and applications in image processing and pattern recognition.

Static Combination Optimization Problem

1. Typical combination of the optimization problems. Since the initial ant colony algorithm to solve the travel quotient problem, researchers have applied it to other typical combinatorial optimization problems: secondary planning problem (Quadratic Assignment Problems) and graph coloring problem. These problems are highly representative of engineering, and the excellent performance of ant colony algorithms on typical combinatorial optimization problems accelerates its development in the field of engineering applications.
2. Application in the field of logistics Some problems in the field of logistics also have the combination optimization problem, mainly the vehicle path problem in the field of logistics distribution. Gambardella first applied the ant colony algorithm to the vehicle path problem. Merle et al. applied the ant colony algorithm to the scheduling problem of resource project constraints, and the experiments showed obvious advantages compared with other heuristics.
3. (3)Institutional isomorphism determination problem. The problem of mechanism isomorphism determination, which is common in the field of mechanical design, is the problem of finding the feature coding value of its neighboring matrix, and uses the powerful search ability of ant colony, which can achieve satisfactory results when the parameter selection is appropriate.
4. Application in the electric power system field. Many optimization problems of power systems are essentially combinatorial optimization problems. Ant colony algorithm was applied to the planning of power distribution networks by Gomez et al. Viacho Giannini Use the ant colony algorithm to solve the constraint power

flow problem, and the experimental results show that the algorithm has high reliability and optimization ability. Fong et al. applied the ant colony algorithm to the optimization of the power plant maintenance plan.

Dynamic Combination Optimization Problem

In the dynamic optimization combination problem, it can be divided into directed connected network routing and unconnected network system routing.

1. Connected network routing. In a directed-connected network, all packets of the same traffic path are transmitted along a common path that is selected by a preliminary set state. Schoonderwerd The first ant colony algorithm applied to the routing problem, then White ACO algorithm used for single to single point and single to multi-point of directional connection network routing, Bonabeau through the introduction of dynamic rule mechanism to improve ant colony algorithm, Dorigo research ant colony algorithm used in high-speed directional connection network system, get the best fair distribution effect.
2. No connected network system routing. As Internet continues to scale, QoS technology is imported on the network to ensure communication quality for real-time services. The purpose of QoS multicast routing is to find the optimal path in the distributed network, which requires starting from the source node, going through all the destination nodes, and achieving the minimum cost service level under all constraints. Ant colony is applied to solve the QoS multicast routing problems including bandwidth, delay, delay jitter, packet loss rate and minimum cost constraint, which is better than the simulated annealing algorithm and genetic algorithm.

Other Application Areas

Other applications of ant colony optimization algorithm include learning fuzzy rule problem, ant automatic planning and design, and artificial neural network, etc.

3.2.7 Related Application of Ant Colony Algorithm and MATLAB Example

Example 3.5 A complete graph of n points has been given, each edge has a length, and the closed loop with the shortest total length passes just once through each vertex.

Separate:

1. Principle of ant group algorithm: Ants release pheromones on the path. When the ant meets a crossroads, it randomly chooses a random path. At the same time, it releases pheromones related to the path length. The concentration of phero-

mones is inversely proportional to the path length. Later, when the ants encounter the intersection again, the ants choose the next city to be transferred based on probability. The path with high pheromone concentration, the pheromone concentration on the optimal path becomes higher and higher, and finally the ant colony finds the optimal feeding path.
2. Ant colony algorithm and TSP problem. Place m ants randomly in multiple cities, let the ants start from the city, n steps (one ant from one city to another city) and then return to the starting city. If the shortest path corresponding to the m path taken by m ants is not the shortest distance of the TSP problem, the process is repeated until the shortest path of the satisfactory TSP problem is found.
3. Path construction. Random proportion rule in the ant colony system: For each ant, the path memory vector records all the city serial numbers that have passed through in the order of visit. Let the current city of ant k be i, and its probability of selecting the city as the next visitor is

$$p_k(i,j) = \begin{cases} \dfrac{[\tau(i,u)]^\alpha [\eta(i,j)]^\beta}{\sum_{j \in J_k(i)} [\tau(i,u)]^\alpha [\eta(i,j)]^\beta}, j \in J_k(i) \\ 0, otherwise \end{cases}$$

Of m: the total number of ants, in the TSP problem, each cycle, each ant out of each path for the problem of TSP candidate solution, m ant a cycle of m path for TSP problem a subset, so the solution subset the greater the algorithm of the global search ability, but the convergence speed of the algorithm. If m is too small, the algorithm can easily fall into the local optimum, premature stagnation.

α: pheromone importance degree factor reflects the degree of the pheromone accumulated on the road between these two cities as the ant selects city j from city i to city j, that is, the strength of the role of the colony in the path search. The larger the alpha value, the more likely the ant will choose the path before, and the randomness of the search path decreases. The smaller the α, the ant colony search range will decrease, and it is easy to fall into the local solution optimal.

β: The important degree factor of enlightening function, the larger the β value, the easier the colony is to choose the local short path. At this time, the convergence rate of the algorithm is accelerated, but the randomness is not high, and it is easy to obtain the local relatively optimal.

τ: a taboo table for the city where the k-th ant has already passed.

ρ: is pheromone volatile factor, regulation $0 < \rho \le 1, 1 - \rho$ for information residue factor, ρ over hours, excessive pheromone on each path, continue to search, affect the invalid path convergence rate of the algorithm, ρ is too large, invalid path can be excluded search, but cannot guarantee the effective path will be abandoned search, affect the optimal value of search, in AS usually set to $\rho = 0.5$.

After 150 iterations, the optimal path is found for 50 cities, and the result is 24722.0443. After execution, the effect diagram is as follows (Figs. 3.14 and 3.15):

3.2 Ant Colony Algorithm

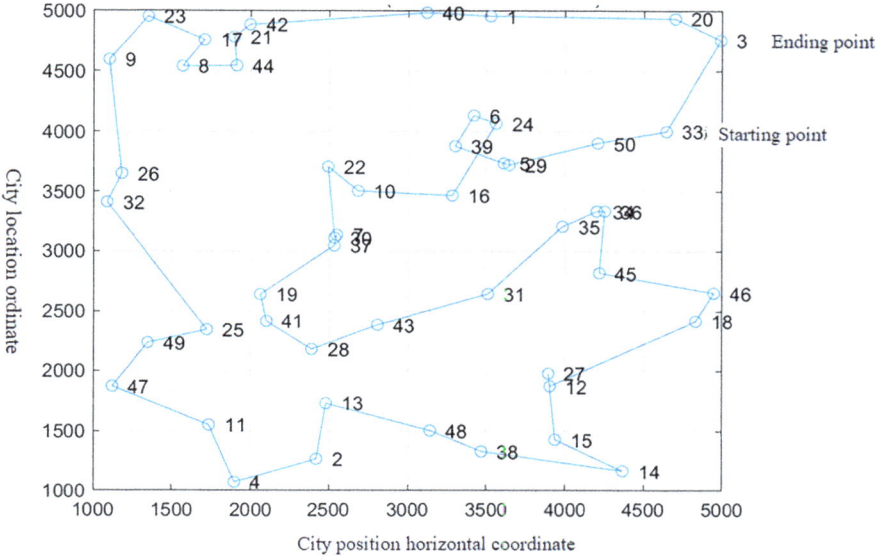

Fig. 3.14 Optimization path of the ant colony algorithm

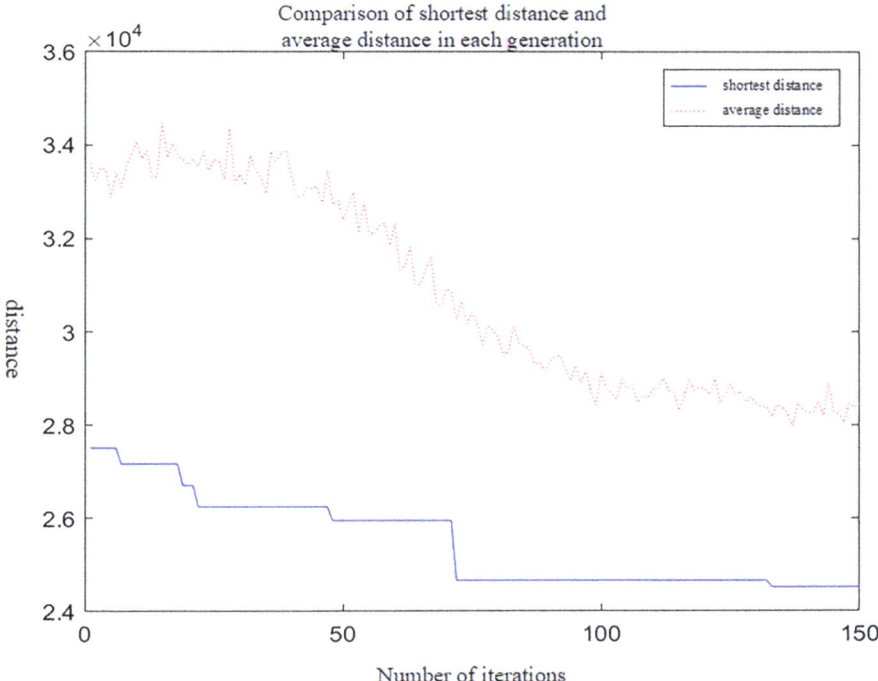

Fig. 3.15 Effect of ant colony algorithm

The development of ant colony algorithm needs a solid theoretical foundation, and the research results in this field are still scarce. Although we can prove the convergence of several types of ant colonies, the current proof of convergence does not explain the computation time needed to find at least one optimal solution. Even if the algorithm can find the optimal solution, the computation time may be astronomical. In addition, the strict mathematics of ant colony algorithm convergence proof, under the stronger probability of convergence conditions, pheromone volatile in colony algorithm of algorithm convergence, colony algorithm dynamic model and according to its dynamic model of the algorithm performance analysis, colony algorithm eventually convergence to the global optimal solution of time complexity also need further research. When using the ant colony algorithm to deal with various problems, what kind of coding scheme, what kind of parameter combination and how to set the artificial pheromone in the algorithm can only be analyzed by specific problems. At present, there is no general, rigorous and scientific model and method. In order to further promote the application and development of ant colony algorithm, the guidance of macro theory is urgently needed.

The Development of the Ant Colony Algorithm Itself

We can consider improving the ant colony algorithm from a new perspective, such as the probabilistic method, multiple group strategy, the improvement of information sharing mechanism between ant colonies, and the introduction of other optimization algorithms. Ant colony algorithm is a population-based method, which has parallelism and can further study the parallelization of the algorithm. Improve the ant colony algorithm itself to control the parameter range to avoid the local optimum due to the parameter being too large or too small.

Comparison and Combination with Other Algorithms

The broadening of the application field of ant colony algorithm should also be cross-studied with other related disciplines. At present, the most successful application of ant colony algorithm is the large-scale combinatorial optimization problem. The next step is to introduce ant colony into more application fields (such as automatic control and machine learning, etc.), and conduct deep crossover research with these related disciplines to further promote the research and development of the algorithm. In addition, the ant colony has strong coupling, easy with other traditional optimization algorithm or heuristic algorithm, but Dr. M. Dorigo pointed out that the ant colony algorithm and distribution estimation algorithm (EDA), graph model and Bayesian network the relationship between the probabilistic method is not clear, this work still need to continue to explore. In the future research, coupling algorithm should be taken as an important research direction, and ant colony and other bionic algorithms should be combined to achieve the effect of learning from each other. Recent achievements have been made in the combination of immune algorithms and genetic algorithms, and the integration of other algorithms needs to be further expanded.

Expand the Application Field of Ant Colony Algorithm

Ant colony algorithms have been introduced in many areas to exert their optimization capabilities. At present, the problem of static combination optimization is widely used. How to improve its application to the dynamic combination optimization problem and the continuous optimization problem is a direction worth exploring.

References

Clerc M, Kennedy J. The particle swarm—explosion, stability, and convergence in a multidimensional complex space. IEEE Transactions on Evolutionary Computation, 2002, 6(1):58-73.

Dorigo M, Caro D G, Stützle T. Ant algorithms. Future Generation Computer Systems,2000,16(8):v-vii.

Eberhart R C, Shi Y. Comparison between genetic algorithms and particle swarm optimization// Evolutionary Programming VII, 7th International Conference, EP98, San Diego, CA, USA, March 25-27, 1998, Proceedings. Springer, Berlin, Heidelberg, 1998.

Kennedy J. Particle swarm optimization. Proc. of 1995 IEEE Int. Conf. Neural Networks, (Perth, Australia), Nov. 27-Dec. 2011, 4(8):1942-1948 vol.4. https://doi.org/10.1007/978-0-387-30164-8_630.

Reynolds, Craig (1987). Flocks, herds and schools: A distributed behavioral model. Association for Computing Machinery. pp. 25–34. https://doi.org/10.1145/37401.37406. ISBN 978-0-89791-227-3.

Trelea I C. The particle swarm optimization algorithm: Convergence analysis and parameter selection. Information Processing Letters, 2003, 85(6):317-325.

Chapter 4
Neural Computing

4.1 BP Neural Network

4.1.1 Concept of the BP Neural Network

In 1986, scientists led by Rumelhart and McClelland put forward the concept of BP neural network. BP neural network is a multi-layer feed-forward neural network trained by the error back propagation algorithm, which is the most widely used neural network.

BP neural network simulates the structure of the brain neural network, and the basic unit of information transmitted by the brain is neurons, there are a large number of neurons in the brain, each neuron is connected to multiple neurons. BP neural network is a simplified biological model, each layer of neural network is composed of neurons, and each neuron alone is equivalent to a perceptron.

The BP neural network is a typical neural network, which is widely used in various classification systems. BP neural network includes two stages: training and use. The training stage is the basis and premise for BP neural network to be put into use, while the use stage itself is a very simple process, that is, to give input. The BP neural network will calculate according to the trained parameters and get the output results (Zhu and Shi 2006).

4.1.2 Model of the BP Neural Network

The BP algorithm includes the forward propagation process of the data flow and the back propagation process of the error signal. The BP neural network uses the learning rules of the fastest descent method to constantly adjust the weights and thresholds of the network through back propagation to minimize the sum of square errors of the network.

The BP Neural Network Structure

As shown in Fig. 4.1, the topology of the BP neural network model includes the input layer, the hidden layer, and the output layer.

Figure 4.2 shows the j-th basic BP neuron, which mimics only the three most basic and most important functions of biological neurons: weighting, sum-finding, and transfer. Where $x_1, x_2, \ldots, x_n, x_2, \ldots, x_i, \ldots, x_n$ represent input from neurons $1,2,\ldots,i,\ldots,n$; $W_{j1}, W_{j2}, \ldots, W_{ji}, \ldots, W_{jn}$ represent the weight of the connection between neuron $1,2,\ldots,i,\ldots,n$ and neuron j, respectively; b_j is the threshold; $f(\cdot)$ is the transfer function; y_j is the output of the j-th neuron.

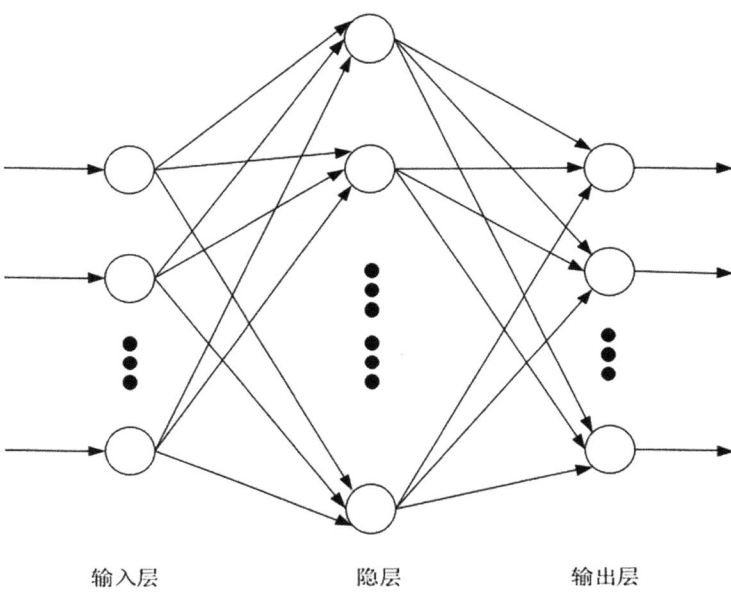

Fig. 4.1 Schematic diagram of the BP neural network structure

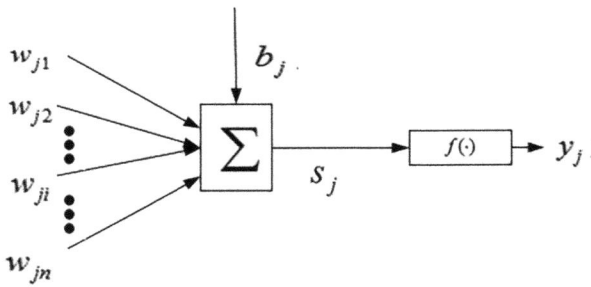

Fig. 4.2 Basic BP neurons

4.1 BP Neural Network

The net input value of the j-th neuron, S_j, is:

$$S_j = \sum_{i=1}^{n} w_{ji} \cdot x_i + b_j = W_j X + b_j \qquad (4.1)$$

If $x_0 = 1$, and $w_{j0} = b_j$, there is:

$$S_j = \sum_{i=0}^{n} w_{ji} \cdot x_i = W_j X \qquad (4.2)$$

Where, $X = [x_0 x_1 x_2 \cdots x_i \cdots x_n]^T$, $W_j = [w_{j0} w_{j1} w_{j2} \cdots w_{ji} \cdots w_{jn}]$.

The net input value S_j is passed through the transfer function $f(\cdot)$ to obtain the output y_j of the j-th neuron:

$$\begin{aligned} y_j &= f(S_j) \\ &= f\left(\sum_{i=0}^{n} w_{ji} \cdot x_i\right) \\ &= F(W_j X) \end{aligned} \qquad (4.3)$$

Where $f(\cdot)$ is a monotonically increasing function and must be bounded, because the signal transmitted by the cell cannot increase infinitely, there must be a maximum.

The Forward Propagation Process

As shown in Fig. 4.3, if the BP network has n input layer nodes, q hidden layer nodes and m output layer nodes, the weight between the input layer and the hidden layer and the hidden layer should be v_{ki}, and the weight between the hidden layer

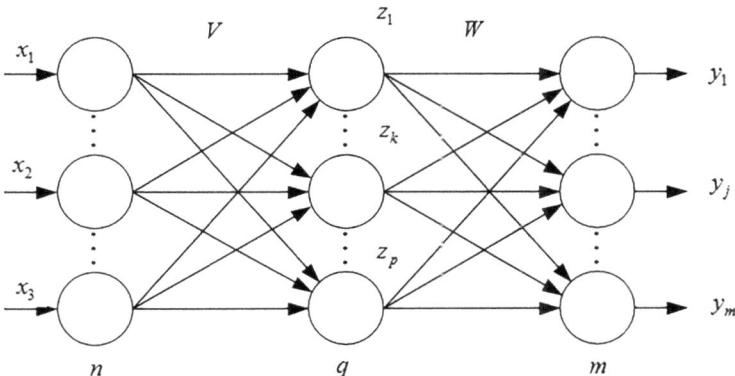

Fig. 4.3 Three-layer BP neural network

and the output layer is w_{jk}. When the hidden layer transfer function and the output layer transfer function are $f_1(\cdot)$ and $f_2(\cdot)$ respectively, the output of the hidden layer node is

$$Z_k = f_1\left(\sum_{i=0}^{n} v_{ki} x_i\right) \qquad (4.4)$$

The output of the output layer node is

$$y_j = f_2\left(\sum_{k=0}^{q} w_{jk} z_k\right) \qquad (4.5)$$

Finally, an approximate map of an n-dimensional space vector to the m-dimensional space.

Error Backpropagation Process

Assuming P learning samples, represented by x^1, x^2, $\cdots x^l$, $\cdots x^P$. The output y_j^l is obtained after the l sample is input to the network, where $j = 1, 2, \cdots m$. Assuming that the expected value of the output of the x^l sample neural network is $\hat{y}_l = \left(\hat{y}_1^l, \hat{y}_2^l \cdots \hat{y}_m^l\right)$, the squared error function is used to obtain the error E_l of the l sample:

$$E_l = \frac{1}{2}\sum_{j=1}^{m}\left(\hat{y}_j^l - y_j^l\right)^2 \qquad (4.6)$$

For the P samples, the global error is

$$E = \frac{1}{2}\sum_{l=1}^{P}\sum_{j=1}^{m}\left(\hat{y}_j^l - y_j^l\right)^2 = \sum_{l=1}^{P} E_l \qquad (4.7)$$

In the next step, the cumulative error BP algorithm is used to adjust the weight w_{jk}, so that the global error E is smaller, given the learning rate η has:

$$\Delta w_{jk} = -\eta \frac{\partial E}{\partial w_{jk}} = -\eta \frac{\partial}{\partial w_{jk}}\left(\sum_{l=1}^{P} E_l\right) = \sum_{l=1}^{P}\left(-\eta \frac{\partial E_l}{\partial w_{jk}}\right) \qquad (4.8)$$

Define the error signal as

$$\delta_{yj} = -\frac{\partial E_l}{\partial S_j} = -\frac{\partial E_l}{\partial y_j} \cdot \frac{\partial y_j}{\partial S_j} \qquad (4.9)$$

4.1 BP Neural Network

where:

$$\frac{\partial E_l}{\partial y_j} = \frac{\partial}{\partial y_j}\left[\frac{1}{2}\sum_{j=1}^{m}\left(\hat{y}_j^l - y_j^l\right)^2\right] = -\sum_{j=1}^{m}\left(\hat{y}_j^l - y_j^l\right) \quad (4.10)$$

$$\frac{\partial y_j}{\partial S_j} = f_2'\left(S_j\right) \quad (4.11)$$

obtain:

$$\frac{\partial E_l}{\partial w_{jk}} = \frac{\partial E_l}{\partial S_j} \cdot \frac{\partial S_j}{\partial w_{jk}} = -\delta_{yj} \cdot z_k = -\sum_{j=1}^{m}\left(\hat{y}_j^l - y_j^l\right)f_2'\left(S_j\right) \cdot z_k \quad (4.12)$$

Thus, the formula (4.8) is adjusted to

$$\Delta w_{jk} = \sum_{l=1}^{P}\sum_{j=1}^{m}\eta\left(\hat{y}_j^l - y_j^l\right)f_2'\left(S_j\right)z_k \quad (4.13)$$

Similar to the adjustment formula for the hidden layer weight:

$$\Delta v_{ki} = \sum_{l=1}^{P}\sum_{j=1}^{m}\eta\left(\hat{y}_j^l - y_j^l\right)f_2'\left(S_j\right)w_{jk}f_1'\left(S_k\right)x_i \quad (4.14)$$

The learning rate $\eta \in (0, 1)$ can control the update step size of each iteration. If the learning rate is set too large, the convergence rate is fast but oscillation is easy to occur, the learning rate is set too small, which requires more iterations, and the convergence will be slow.

It is difficult to fully understand the backpropagation algorithm by using obscure formulas, and we can better understand the algorithm by using a detailed example. In order to be as simple and easy to understand as possible, there is no calculation of complex neural network back propagation process, complex calculation can be completed by computer, as long as understanding the simple neural network back propagation process can be inferred from one example, using the powerful computer to realize the calculation of complex network.

Example 4.1 Figure 4.4 illustrates a basic three-layer neural network with two inputs and one output. The network comprises an input layer (x_1, x_2), an output layer y, and a hidden layer (z_1, z_2). There are six weights—v_{11}, v_{12}, v_{21}, v_{22}, w_{11}, w_{12}—connecting the input layer to the hidden layer and the hidden layer to the output layer, respectively, facilitating the training process of this network.

Solution After understanding the network structure, began to consider the initial parameters of the network, six weights can be generated by way of random initial weight, both Python or MATLAB can use the corresponding function to generate

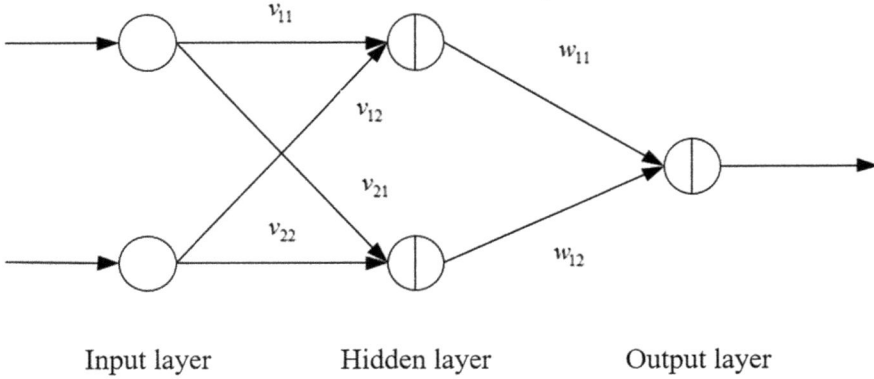

Fig. 4.4 3 layer neural network

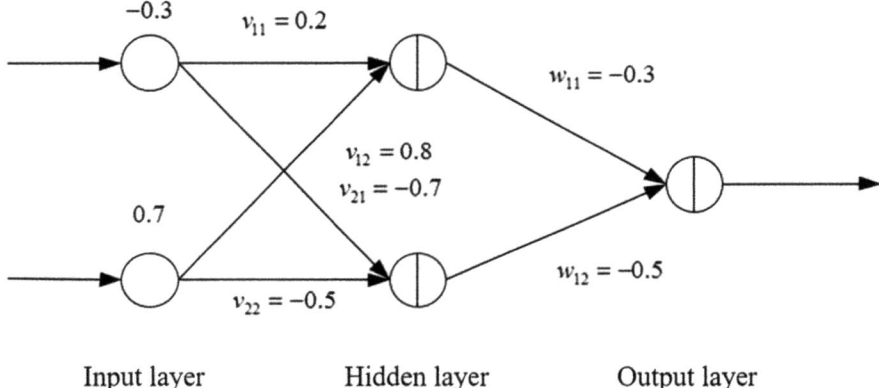

Fig. 4.5 Neural network iterations

random number, but it is worth noting that the range of random number cannot without limit, general requirement range is (1, −1), activation function using Sigmoid function, the learning rate is set to 0.6.

Now train the network, assuming the first data input is (−0.3, 0.7) and the target output is 0.1. First, six initial weights were randomly obtained:

$$v_{11} = 0.2, v_{12} = 0.8, v_{21} = -0.7, v_{22} = -0.5, w_{11} = -0.3, w_{12} = -0.5$$

In order to think clearer, it is necessary to pay attention to the changes of network parameters all the time. Figure 4.5 shows the current network, and marks the weight parameters of the current network and the current values of each node.

Weighted sums of input layer to hidden layer nodes:

$$-0.3 \times 0.2 + 0.7 \times 0.8 = 0.5$$

4.1 BP Neural Network

$$-0.3 \times (-0.7) + 0.7 \times (-0.5) = -0.14$$

Perform the Sigmoid activation function:

$$\text{logsig}(0.5) = \frac{1}{1+e^{-0.5}} = 0.622$$

$$\text{logsig}(-0.14) = \frac{1}{1+e^{0.14}} = 0.535$$

Figure 4.6 shows the current case of the neural network.
Weighted summation of hidden layer to output layer nodes:

$$0.622 \times (-0.3) + 0.535 \times (-0.5) = -0.454$$

Execute the Sigmoid activation function of the output layer:

$$\text{logsig}(-0.454) = \frac{1}{1+e^{0.454}} = 0.388$$

At this point, the forward propagation of the signal is completed. Figure 4.7 shows the current situation of the neural network.

The output result is 0.388, which is inconsistent with the expected output 0.1. The error and residual of the sample are calculated below. The residual here refers to the partial derivative of the error. Where, the error is

$$(0.388 - 0.1)^2 = 0.083$$

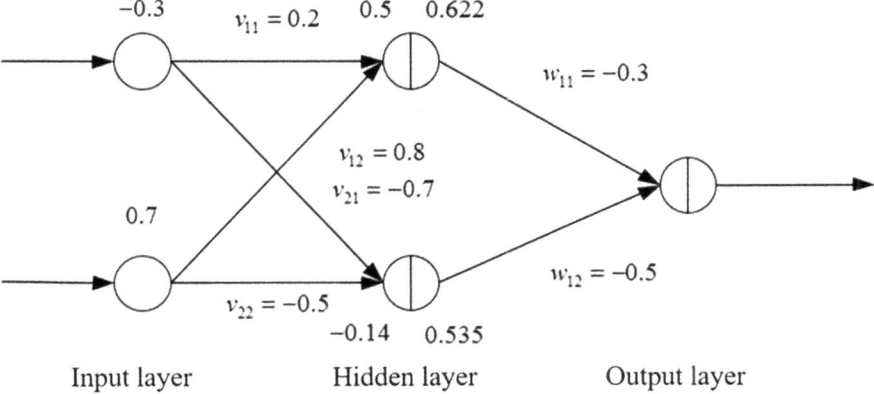

Fig. 4.6 The second step of neural network iteration

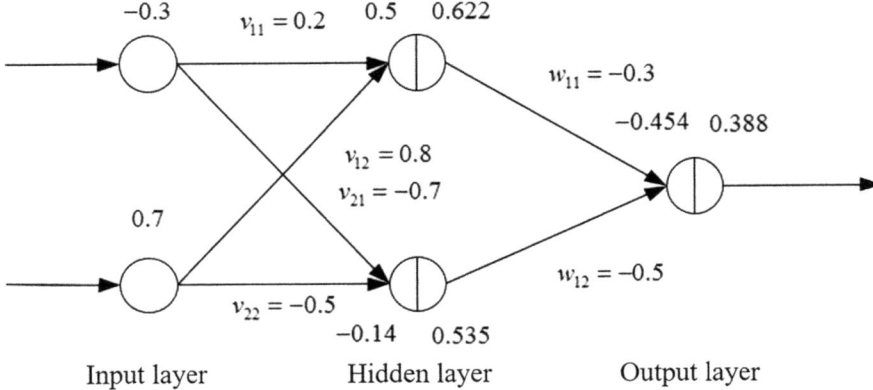

Fig. 4.7 The third step of neural network iteration

The residual is

$$-(0.388-0.1) \times 0.388 \times (1-0.388) = -0.068$$

Next, backpropagation begins, weighted summing the output layer to the hidden layer:

$$-0.068 \times (-0.3) = 0.020$$

$$-0.068 \times (-0.5) = 0.034$$

Then find the residual of the hidden layer:

$$0.020 \times 0.622 \times (1-0.622) = 0.005$$

$$0.034 \times 0.535 \times (1-0.535) = 0.008$$

Figure 4.8 shows the backpropagation of the error of the current BP neural network. The residual error of the output layer is −0.068, and the residual error of the hidden layer is 0.005 and 0.008 respectively.

Now update the weight between the input layer and the hidden layer, set the learning rate of 0.6, and the weight from top to bottom should be updated at

$$0.3 \times 0.005 \times 0.6 = 0.0009$$

$$0.3 \times 0.008 \times 0.6 = 0.0014$$

$$-0.7 \times 0.005 \times 0.6 = -0.0021$$

$$-0.7 \times 0.008 \times 0.6 = -0.0034$$

4.1 BP Neural Network

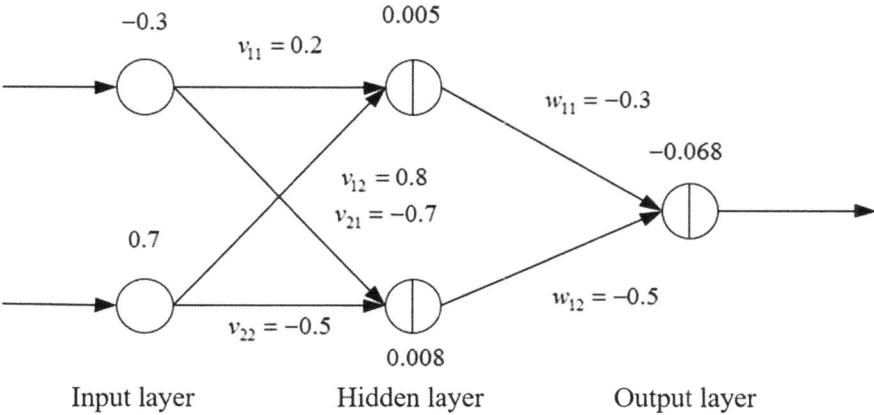

Fig. 4.8 Backpropagation

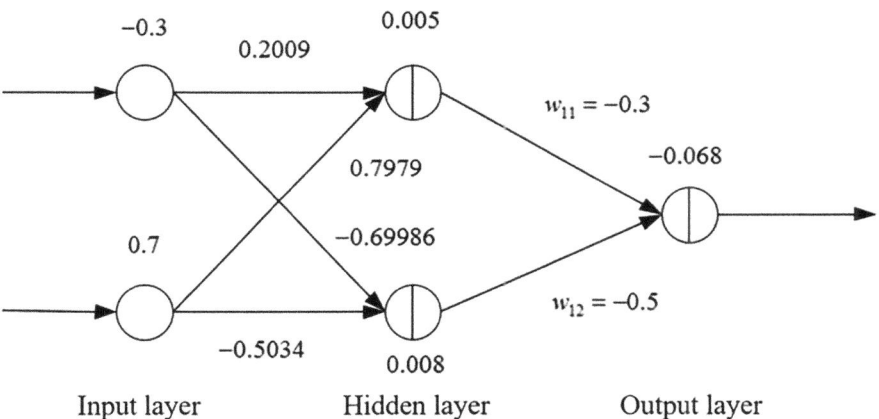

Fig. 4.9 Weight variation

The updated weight is given as follows

$$0.2 + 0.0009 = 0.2009$$

$$-0.7 + 0.00014 = -0.69986$$

$$0.8 - 0.0021 = 0.7979$$

$$-0.5 - 0.0034 = -0.5034$$

As shown in Fig. 4.9, it can be found that the weight from the input layer to the hidden layer has changed. Here we can really understand that the essence of training is to constantly change the connection weight between each node.

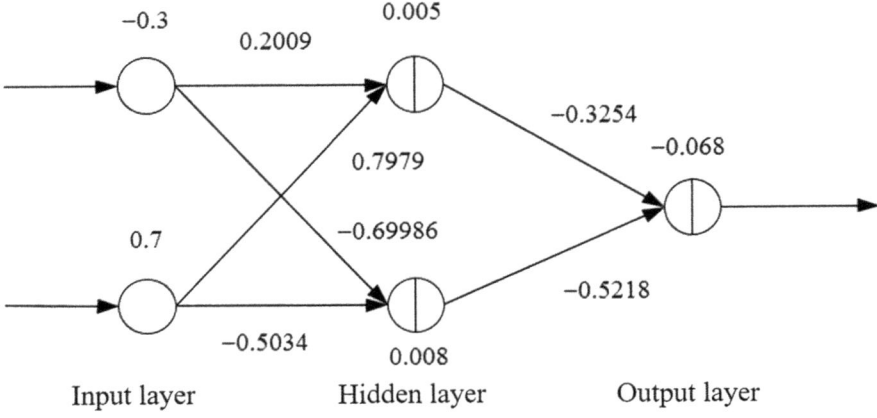

Fig. 4.10 A complete neural network learning process

Update the weight between the hidden layer and the output layer, and the learning rate is still 0.6:

$$0.622 \times (-0.068) \times 0.6 = -0.0254$$

$$0.535 \times (-0.068) \times 0.6 = -0.0218$$

The updated weight is:

$$-0.3 - 0.0254 = -0.3254$$

$$-0.5 - 0.0218 = -0.5218$$

As shown in Fig. 4.10, the six weights of the BP neural network have all changed. So far, a "learning" has been completed, and a gradient descent and weight update have been essentially completed. Through Example 4.1, we can specifically understand how the BP neural network works.

4.1.3 Characteristics of the BP Neural Network

The BP neural network has the following advantages.
  Nonlinear mapping capability. BP neural network essentially realizes a mapping function from input to output. Mathematical theory proves that 3-layer neural network can approximate any nonlinear continuous function with arbitrary precision. This makes it especially suitable for solving problems with complex internal mechanism, that is, BP neural network has strong nonlinear mapping ability.

4.1 BP Neural Network

Self-learning and adaptive ability. During training, BP neural network can automatically extract the "reasonable rules" between input and output data through learning, and adaptively memorize the learning content in the weight of the network, that is, BP neural network has a high degree of self-learning and adaptive ability.

Generalization ability. The so-called generalization ability means that when designing the mode classifier, it should not only consider the network to ensure the correct classification of the required classified objects, but also care whether the network can correctly classify the unseen mode or patterns with noise pollution after training. That is, the BP neural network has the ability to apply the learning results to the new knowledge.

Fault tolerance. The BP neural network will not greatly affect the global training results after the destruction of the local or part of its neurons, that is, the system can still work normally even when it is locally damaged. That is, the BP neural network has a certain fault-tolerance ability.

In view of these advantages of BP neural network, many researchers at home and abroad have studied it, and used the network to solve many application problems. However, with the gradual expansion of the application scope, BP neural network has also exposed more and more shortcomings and deficiencies.

Local minimization problem. From a mathematical point of view, the traditional BP neural network is a kind of local search optimization method, it is to solve a complex nonlinearization problem, the weight of the network is gradually adjusted along the local improvement direction, this will make the algorithm into local extreme value, weight convergence to local minimum point, leading to network training failure. BP neural network is very sensitive to the initial network weight, and initializes the network with different weights, which tends to converge to different local minima, which is also the fundamental reason why many scholars get different results for each training.

The convergence rate of the BP neural network algorithm is slow. Since BP neural network algorithm is essentially gradient descent method, the objective function to be optimized is very complex, so there will inevitably be "zigzag phenomenon", which makes BP algorithm inefficient; and because the optimized objective function is very complex, it will inevitably appear flat zone in the neuron output close to 0 or 1, where the weight error changes very little, making the training process almost stop. In BP neural network model, in order to make the network perform BP algorithm, the traditional one-dimensional search method cannot be used to find the step size of each iteration, but the update rule of the step size must be given to the network in advance. This method will also cause the inefficiency of the algorithm. All of these lead to the slow convergence rate of BP neural network algorithm.

The choice of BP neural network structure is different. So far, there is no unified and complete theoretical guidance for the selection of BP neural network structure, and it can only be selected by experience. The network structure selection is too large, the training efficiency is inefficient, and the overfitting phenomenon may occur, resulting in low network performance and fault tolerance decrease; if the selection is too small, the network may not converge. The structure of the network directly affects the approximation ability and extension nature of the network. Therefore, how to select the appropriate network structure in the application is an important issue.

The contradiction between application examples and network scale. It is difficult to solve the contradiction between the instance size and the network size of the application, which involves the relationship between the possibility and the feasibility of the network capacity, namely the learning complexity problem.

The contradiction between the predictive ability and the training ability of the BP neural network. Predictive ability is also called generalization ability or promotion ability, and training ability is also called approximation ability or learning ability. In general, when the training ability is poor, the prediction ability is also poor, and to a certain extent, with the improvement of the training ability, the prediction ability will be improved. However, this trend is not fixed, it has a limit, when this limit is reached, as the training ability improves, the prediction ability will decline, that is, the so-called overfitting phenomenon appears. The reason for this phenomenon is that the network has learned too many sample details, and the learned model can no longer reflect the rules of the sample inclusion. Therefore, how to grasp the degree of learning and solve the contradiction between network prediction ability and training ability is also an important research content of BP neural network.

Sample-dependence problem of the BP neural network. The approximation and generalization ability of the network model are closely related to the typicality of the learning sample, and it is a very difficult problem to select the typical sample instances from the problem to form the training set.

4.1.4 Related Applications of BP Neural Network and MATLAB Examples

BP neural network is widely used in various classification systems as well as prediction systems. The following example is a simple application (Shi et al. 2009).

Example 4.2 Table 4.1 shows the sales situation of a drug. Now a 3-layer BP neural network is built to predict the sales of drugs. There are 3 nodes in the input layer, the number of hidden layer nodes is 5, the activation function of the hidden layer is tansig; the number of nodes in the output layer is 1, and the activation function of the output layer is logsig to predict the sales volume of drugs. Prediction method using rolling forecast way, namely with the first 3 months of sales to predict the fourth month sales, such as in 1, 2, 3 sales to forecast the fourth month sales, with

Table 4.1 Sales situation of a certain drug

Month	1	2	3	4	5	6
Sales	2056	2395	2600	2298	1634	1600
Month	7	8	9	10	11	12
Sales	1873	1478	1900	1500	2046	1556

4.1 BP Neural Network

2, 3, 4 input sales for the forecast of the fifth month sales, so repeatedly until meet the prediction accuracy requirements.

Solution The MATLAB procedure is as follows:

```
% Sales of every three months are normalized as input
P=[0.5152    0.8173    1.0000;
   0.8173    1.0000    0.7308;
   1.0000    0.7308    0.1390;
   0.7308    0.1390    0.1087;
   0.1390    0.1087    0.3520;
   0.1087    0.3520    0.0000;]';
%The sales volume of the fourth month is normalized as the target vector
T=[0.7308 0.1390 0.1087 0.3520 0.0000 0.3761];
% Create a BP neural network, the value range of each input vector is [0,1], 5 hidden layer neurons, an output layer neurons%
hidden layer activation function tansig, output layer activation function logsig, training function is gradient descent function.
   net=newff([0 1;0 1;0 1], [5, 1],{'tansig','logsig'},'traingd');
   net.trainParam.epochs=15000;
   net.trainParam.goal=0.01;
   LP.lr=0.1; %Set the learning rate to 0.1
   net=train(net,P,T);
```

The obtained prediction effect has some error with the actual one, which can be further reduced by increasing the number of running steps and improving the preset error accuracy.

4.1.5 Algorithmic Improvement of the BP Neural Network

The theory of BP algorithm has the advantages of reliable basis, rigorous deduction process, high accuracy and good versatility, but the standard BP algorithm has the following disadvantages: slow convergence rate; easy to fall into local minimum; it is difficult to determine the number of hidden layers and the number of hidden layer nodes. In practical application, the BP algorithm is difficult to perform, so many improved algorithms have emerged.

Improve the BP Algorithm by the Momentum Method

In essence, the standard BP algorithm is a simple static optimization method of the fastest descent. When correcting W(K), it only corrects according to the negative gradient direction in step K, without taking into account the previous accumulated experience, that is, the gradient direction of the previous moment, so that the learning process can oscillate and convergence is slow. The specific practice of the momentum method weight adjustment algorithm is: add part of the last weight adjustment to the weight adjustment calculated according to the error, as the actual weight adjustment of this time, namely

$$\Delta W(n) = -\eta \nabla E(n) + \alpha \Delta W(n-1) \quad (4.15)$$

where α is the momentum coefficient, usually $0 < \alpha < 0.9$; η is the learning rate, ranging from 0.001 to 10. The momentum factor added to this method is actually equivalent to the damping term, which reduces the oscillation trend during learning and improves convergence. The momentum method reduces the sensitivity of the network to the local details of the error surface and effectively prevents the network from falling into the local minimum.

An important reason for the slow convergence of standard BP algorithm is that the learning rate is not selected properly, the learning rate is selected too small, and the convergence is too slow. If the learning rate is chosen too large, it may be corrected too much, resulting in oscillation or even divergence. In this case, the adaptive method shown in Fig. 4.11 can be used to adjust the learning rate.

The basic guiding principle of adjustment is: in the case of learning convergence, increase η to shorten the learning time; when η is too large to cause the convergence, to reduce η in time until convergence.

When using momentum method, BP algorithm can find better solution. When using adaptive learning rate method, BP algorithm can shorten the training time.

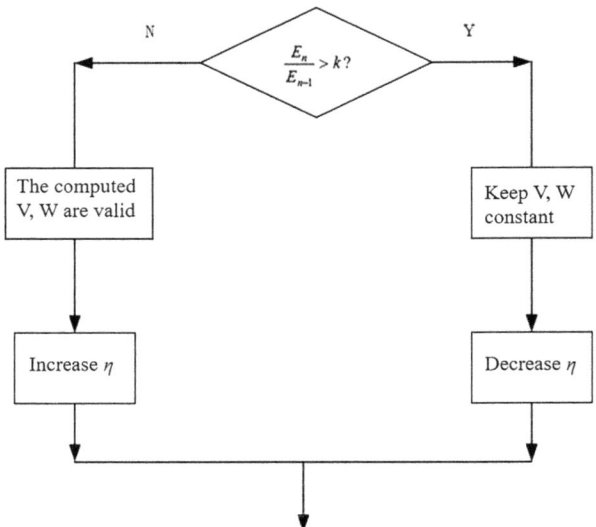

Fig. 4.11 Adaptive learning

4.1 BP Neural Network

The momentum-adaptive learning rate adjustment algorithm is obtained by combining the above two methods.

The L-M (Levenberg-Marquardt) algorithm is much faster than the aforementioned BP algorithms using gradient descent methods. But for complex problems, this method requires considerable storage space with an incremental equation of

$$\left(J^T W J + \lambda I\right)\delta x = -J^T W \Delta z \tag{4.16}$$

The L-M algorithm emphasizes the selection of the parameter λ. When λ approaches infinity, the L-M algorithm is equivalent to the gradient descent method; whe λ approaches zero, it approximates the Gauss-Newton method. It is recommended to initially set a value for λ_0 and a coefficient $v > 1$. Subsequently, update λ using $\lambda = \lambda_0$ and $\lambda = \lambda_0/v$, respectively, and compute the cost function. If the cost function increases, continue multiplying by v until a value is found that reduces the cost function.

Improvement of the Loss Function

In general, the activation function commonly used in BP network is Sigmod function. As can be seen from the Sigmod image, during the training of the neural network, when the output of the last layer is close to 0 or 1, the gradient will disappear, and the learning speed of the neural network will slow down. At this point, define the cross-entropy cost function:

$$C = \frac{1}{n}\left(\sum_{n_x} -y \ln a^L + (1-y)\ln\left(1-a^L\right)\right) \tag{4.17}$$

$$\frac{\partial C}{\partial w_j^l} = \frac{1}{n}\sum_{n_x}\left(-\frac{y}{a^L} + \frac{1-y}{1-a^L}\right)\frac{\partial \sigma\left(z^L\right)}{\partial w_j^l} a_j^{l-1} \tag{4.18}$$

where, $w_j^l = \left(w_{j,1}^l, w_{j,2}^l, \ldots, w_{j,l-1}^l\right)$ represents the weight of each neuron in layer $l-1$ to the j-th neuron in layer l; a_j represents the output of layer j, expressed as $a^j = \sigma(z^{j-1})$, $j = 2, 3, \ldots L$, L is the number of layers of the neural network; $z^j = w_j' a^{j-1} + b^j$, the formula (4.18) can be rewritten as:

$$\frac{\partial C}{\partial w_j^l} = \frac{1}{n}\sum_{n_x}\left(-\frac{y}{a^L} + \frac{1-y}{1-a^L}\right)\frac{\partial \sigma\left(z^L\right)}{\partial z^l}\frac{\partial z^l}{\partial w_j^l} a_j^{l-1} \tag{4.19}$$

can be obtained

$$\frac{\partial C}{\partial w_j^l} = \frac{1}{n}\sum_{n_x}\left(a^L - y\right)\frac{\partial \sigma\left(z^L\right)}{\partial z^l}\frac{\partial z^l}{\partial w_j^l} a_j^{l-1} \tag{4.20}$$

Formula (4.20) solves the disadvantage of slow learning when the output layer has a large deviation from the sample result, and it is independent of the σ'(zL) of the last layer, so the disappearance of the gradient can be avoided.

The Bayesian Regularization Algorithm

In addition to the above methods, Bayesian regularization algorithm can also optimize BP neural network, and using Bayesian regularization algorithm can improve the generalization ability of BP network.

Example 4.3 Two training methods, namely, L-M optimization algorithm (trainlm) and Bayesian regularization algorithm (trainbr), are used to train the BP network to fit a certain sinusoidal sample data with attached white noise.

Solution The training results are shown in Figs. 4.12 and 4.13 respectively, and the MATLAB implementation code is as follows.

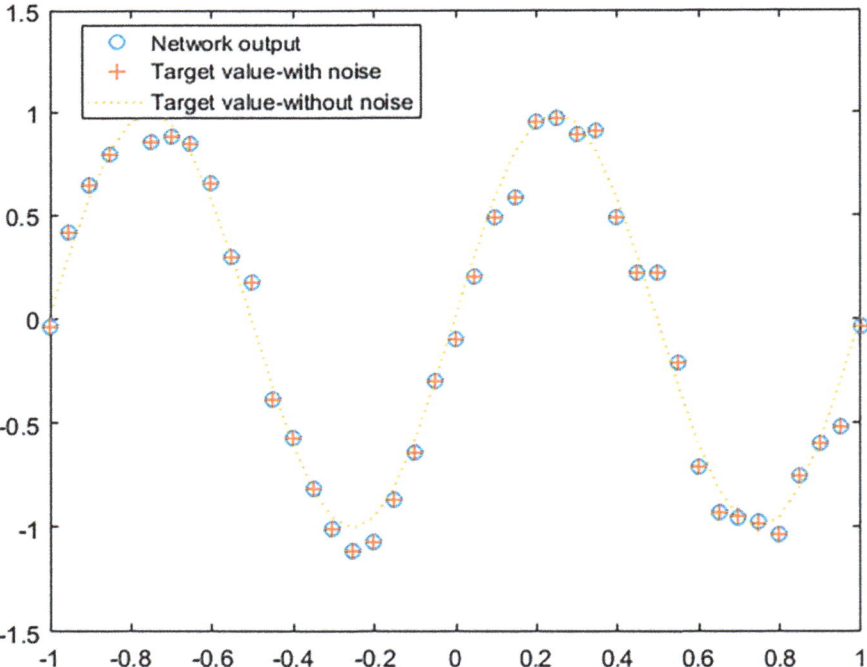

Fig. 4.12 Fitting results using the L-M optimization algorithm

4.1 BP Neural Network

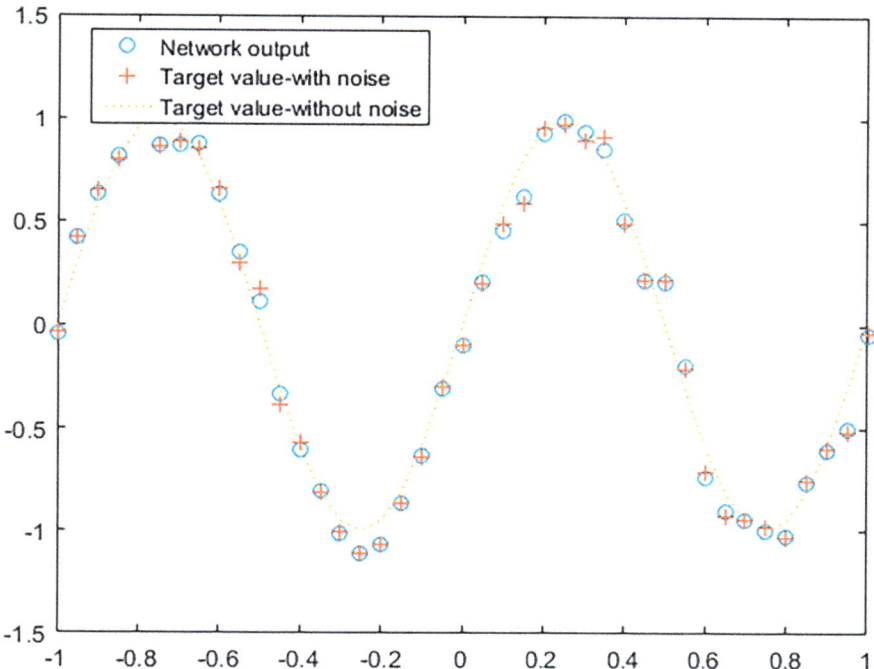

Fig. 4.13 The fitting results obtained using a Bayesian regularized optimization algorithm

```
% MATLAB statement generation:
    % Input vector: P= [-1:0.05:1];
    % Target vector: randn (seed, 78341223);
    % T =sin (2 * pi * P) + 0.1 * randn (size (P));
    % MATLAB procedure is as follows:
    close all
    clear all
    clc
    % NEWFF indicates generating a new forward neural network
    % TRAIN means training the BP neural network, and SIM means
simulating the BP neural network
    % Defines the training sample vector
    P = [-1:0.05:1];                                        % P
is the input vector
    randn('seed',78341223);
    T = sin(2*pi*P)+0.1*randn(size(P));                    % T is the
target vector
    net=newff(minmax(P),[20,1],{'tansig','purelin'});
    % Create a new forward neural network
    disp('1.L-M optimization algorithm TRAINLM'); disp('2.Bayesian
```

```
regularization algorithm TRAINBR');
   choice=input('Please select the training algorithm(1,2):');
   if(choice==1)
       net.trainFcn='trainlm';              % Using the L-M
optimization algorithm
   net.trainParam.epochs = 500;
       net.trainParam.goal = 1e-6;
       net=init(net);                       % Reinitialized
       pause;
   elseif(choice==2)                        % Use the Bayesian
regularization algorithm
       net.trainFcn='trainbr';
       net.trainParam.epochs = 500;         % Set the training
parameters
       net = init(net);                     % Reinitialized
       pause;
   end
   [net,tr]=train(net,P,T);     % The corresponding algorithm
was called to train the BP network
   A = sim(net,P);                          % Simulations of the BP
network were performed
   E = T - A;                                         % Calculate
the simulation error
   MSE=mse(E)
   figure
   plot(P,A,'o',P,T,'+',P,sin(2*pi*P),':'); % Draw the matching
result curve
   legend('Network output','Target value-with noise','Target
value-without noise')
```

It can be seen that the neural network trained by the trainlm function realizes the sample data points, while the neural network trained by the trainbr function is insensitive to noise and is robust.

Genetic Algorithm Optimization Method

Previously, when using BP network, the weight and threshold can be assigned to the network before training the network. In other words, the weight and threshold value of the neural network is assigned how. Of course, the result of the value assignment will definitely affect the final prediction performance of the neural network. However, the goal is to make the final prediction performance of the neural network the best, that is to say, to find the best weight and threshold. At this time, intelligent optimization algorithm can be used to search the weight and threshold. At present, many intelligent optimization algorithms have appeared, such as using genetic algorithm to optimize BP network.

4.1 BP Neural Network

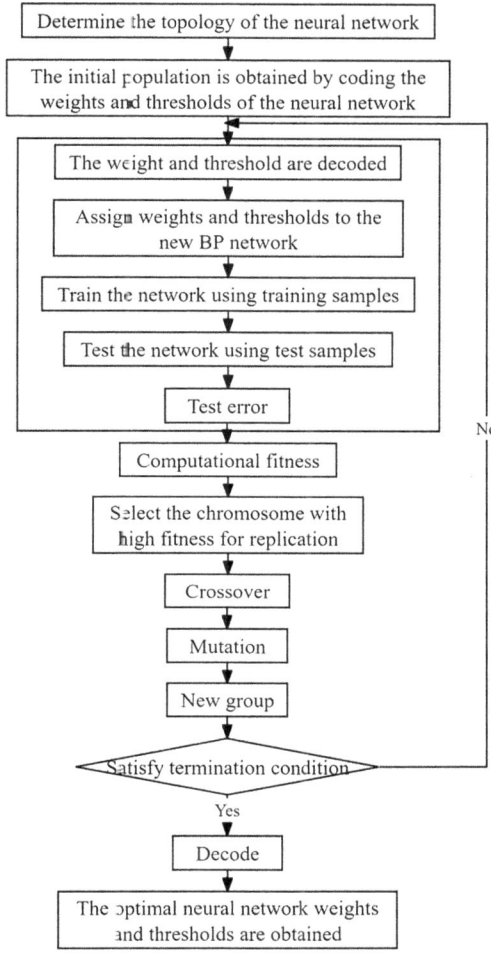

Fig. 4.14 BP neural network process improved by genetic algorithm

The weights and thresholds were optimized using a genetic algorithm. Because the weights and threshold are real numbers, the encoding method is binary coding. With binary encoding, it is more convenient to use the Sheffield genetic algorithm toolbox. Figure 4.14 shows the BP neural network flow of the genetic algorithm.

The main procedure is as follows:

```
clc
grid onxlabel('Genetic algebra')
ylabel('Variation of error')
title('Evolutionary process')
bestX=trace(1:end-1,end);
bestErr=trace(end,end);
fprintf(['Optimal initial weights and thresholds:\
nX=',num2str(bestX),
```

```
'\nMinimum errorerr=',num2str(bestErr),'\n'])
%% Compare training & testing before and after optimization
callbackfun
```

The code for finding the population objective function value is as follows:

```
function Obj=Objfun(X,P,T,hiddennum,P_test,T_test)
%% It is used to solve the target value of each individual in the population separately
%% Input
% X:Initial weights and thresholds for all individuals
% P:Training sample input
% T:Training sample output
% hiddennum:Number of hidden layer neurons
% P_test:Test sample input
% T_test:Test sample expected output
%% Output
% Obj:The norm of prediction error for all individual prediction samples
[M,N]=size(X);
Obj=zeros(M,1);
for i=1:M
Obj(i)=BPfun(X(i,:),P,T,hiddennum,P_test,T_test);
end
```

Conventional selection, crossover, mutation, reinsertion and other operations, you can use the toolbox functions to achieve the above operations, the code is as follows:

```
FitnV=ranking(ObjV);                        %Assign fitness values
SelCh=select('sus',Chrom,FitnV,GGAP);       %Selection
SelCh=recombin('xovsp',SelCh,px);           %Crossover
SelCh=mut(SelCh,pm);                        %Mutation
X=bs2rv(SelCh,FieldD);         %Decimal conversion of child individuals
ObjVSel=Objfun(X,P,T,hiddennum,P_test,T_test);
%Calculate the objective function value of the child
[Chrom,ObjV]=reins(Chrom,SelCh,1,1,ObjV,ObjVSel);
%The offspring are reinserted into the parent generation to get a new population
X=bs2rv(Chrom,FieldD);
```

An individual consists of four parts: the weight from the input layer to the hidden layer, the threshold value of the hidden layer, the weight from the hidden layer to the output layer, and the threshold value of the output layer. The code for finding the objective function value of an individual is as follows:

4.1 BP Neural Network

```
function err=BPfun(x,P,T,hiddennum.P_test,T_test)
%% Training & Testing BP networks
%% Input
% x:The initial weight and threshold of an individual
% P:Training sample input
% T:Training sample output
% hiddennum:Number of hidden layer neurons
% P_test:Test sample input
% T_test:Test sample expected output
%% Output
% err:The norm of the prediction error of the prediction sample
   inputnum=size(P,1);                      % Number of neurons in the input layer
   outputnum=size(T,1);                     % Number of neurons in the output layer
   %% New BP network
   net=newff(minmax(P),[hiddennum,outputnum],{'tansig','logsig'},'trainlm');
   %% Set the network parameters: the training number is 1000, the training target is 0.01, and the learning rate is 0.1
   net.trainParam.epochs=1000;
   net.trainParam.goal=0.01;
   LP.lr=0.1;
   net.trainParam.show=NaN;
   %% Initial weight and threshold of BP neural network
   w1num=inputnum*hiddennum;                %The number of weights from the input layer to the hidden layer
   w2num=outputnum*hiddennum;               %The number of weights from the hidden layer to the output layer
   w1=x(1:w1num);                           %Weights from the initial input layer to the hidden layer
   B1=x(w1num+1:w1num+hiddennum);           %Initial hidden layer threshold
   w2=x(w1num+hiddennum+1:w1num+hiddennum+w2num);
   %Threshold from the initial hidden layer to the output layer
   B2=x(w1num+hiddennum+w2num+1:w1num+hiddennum+w2num+outputnum);
   %Output layer threshold
   net.iw{1,1}=reshape(w1,hiddennum,inputnum);
   net.lw{2,1}=reshape(w2,outputnum,hiddennum);
   net.b{1}=reshape(B1,hiddennum,1);
   net.b{2}=reshape(B2,outputnum,1);
   %% Training network
   net=train(net,P,T);
   %% Test network
   Y=sim(net,P_test);
   err=norm(Y-T_test);
```

4.2 Deep Neural Network

4.2.1 *Concept of the Deep Neural Network*

Although neural networks were introduced in the 1940s, it was not until the late 1980s that the first practical application began: LeNet, which recognizes handwritten numbers. This system is widely used in the check number recognition. Since 2010, the application of deep neural networks (DNN) has exploded.

In the 1980s, Rumelhart, Williams, Hinton, and others invented the Multi-Layer Perceptron (MLP), which solved the problem of not being able to simulate the heterogeneous logic before, and at the same time, more layers also made the network more capable of portraying the complexity of real-world situations. As the name suggests, a multi-layer perceptron is a perceptron with multiple hidden layers.

At the same time, scientists have found that the number of layers of a neural network directly determines its ability to portray reality, as the number of layers of the neural network deepens, the optimization function is more and more likely to fall into the local optimal solution, and this "trap" is more and more deviated from the true global optimum. The performance of the deep network trained with limited data is not as good as that of the shallow network. At the same time, another problem cannot be ignored is that as the number of network layers increases, the phenomenon of gradient disappearance becomes more serious.

In 2006, Hinton used the pre-training method to alleviate the problem of local optimal solution, and pushed the hidden layer to 7 layers, the neural network really has the meaning of "depth", which opened the deep learning craze. There is no fixed definition of "depth". However, when there are too many layers, the gradient disappears. In order to overcome the gradient disappearance, ReLU, maxout and other transfer functions replace the sigmoid, forming the basic form of deep neural networks today.

Deep neural networks are often referred to as deep learning, and the huge success of deep learning around 2010 is mainly due to three factors.

The huge amount of information required to train the network. Learning an effective representation requires a large amount of training data. Currently, Facebook receives more than 350 million images per day, Walmart generates 2.5 petabytes of user data per hour, and YouTube uploads nearly 300 hours of video per minute. Therefore, cloud service providers have a huge amount of data that can be used for algorithm training.

Adequate computing resources. Advances in semiconductors and computer architecture have provided sufficient computing power to make it possible to train algorithms in a reasonable amount of time.

The evolution of algorithmic techniques has greatly improved the accuracy and broadened the scope of DNN applications. Early applications of DNNs opened the door to algorithmic development, which inspired the development of a number of deep learning frameworks (most of which are open source) and made DNN networks easily accessible to a wide range of researchers and practitioners.

4.2 Deep Neural Network

The ImageNet challenge is a great example of machine learning success. The challenge involves competing in several different directions. The first direction is image classification, where an algorithm given an image must recognize the content in the image. The training set consists of 1.2 million images, each labeled with one of 1000 object categories contained in the image. The algorithm must then accurately recognize the images in the test set.

Based on the performance of the best participants in each year of the ImageNet challenge over the years, the initial algorithm had an error rate of 25% or more. 2012 saw a team from the University of Toronto reduce the error rate by about 10%, using the high computational power of GPUs and the deep neural network approach, AlexNet. Their achievement led to the popularization and continuous improvement of deep learning style algorithms.

The number of teams using deep learning methods in the ImageNet challenge and the number of participants using GPU computing has increased accordingly. In 2012, only four teams used Gpus, but in 2014, almost all of them used Gpus. This reflects a complete shift from traditional computer vision approaches to deep learning.

In 2015, the ImageNet award-winning ResNet surpassed the human-level accuracy (Top-5 error rate of less than 5%), reducing the error rate to less than 3%. The current focus of DNNs is not so much on accuracy improvement, but rather on other more challenging directions, such as object detection and localization. These successes are clearly a reason for the wide range of applications of DNNs (Goodfellow et al. 2017).

4.2.2 Model of Deep Neural Network

To understand the structure of a deep neural network, we must first understand the perceptron. The perceptron was invented in the 1950s to the 1960s by scientist Frank Rosenblatt, who was inspired by earlier work from Warren McCulloch and Walter Pitts. Today, DNN usually uses other types of artificial neuronal models, and in many of the latest work on neural networks, a neuronal models called Sigmoid neurons (Sigmoid neuron). Figure 4.15 is a schematic diagram of the structure of perceptron. It can be seen that the model of perceptron has several inputs and one output.

Since the neuron activation function is sign() function, which can only get the output result of 1 or −1, the perceptron model can only realize the binary

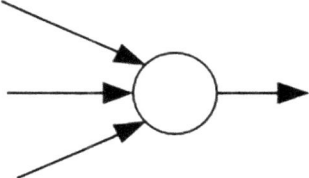

Fig. 4.15 Schematic diagram of the perceptron neural network structure

classification, which is powerless for more complex cases. When the network contains more than one perceptron and connected to each other, a deep neural network model is formed, which can be flexibly applied to classification, regression, downscaling and clustering.

Simple Deep Neural Network

Figure 4.16 shows a simple deep neural network structure schematic, in fact, deep neural networks can have more than a dozen layers or even more layers, DNN internal neural network can be divided into the input layer, the hidden layer and the output layer, generally the first layer is the input layer, the last layer is the output layer, and the intermediate layers are all hidden layers. The layers are fully connected to each other, i.e. any neuron in layer i must be connected to any neuron in layer i + 1.

Example 4.4 Use MATLAB to construct a DNN for training and testing.

Solution MATLAB deep learning toolbox is used to realize DNN, the use of deep learning toolbox is similar to Keras, the training model can be summarized in four steps: define the training data → define the neural network model → configure the learning process → train the model.

Training data. The training data is the simulation signal $y = \cos(2\pi wt + \phi)$, and then the noise is added by using the awgn() function. The input value is a section of the delay signal of y, 2000 points in total, which is denoted as x; the output value is a point of the signal of y itself; there is a corresponding relationship between the two. Suppose y is [12000,1], then the corresponding length of x is [14000,1].

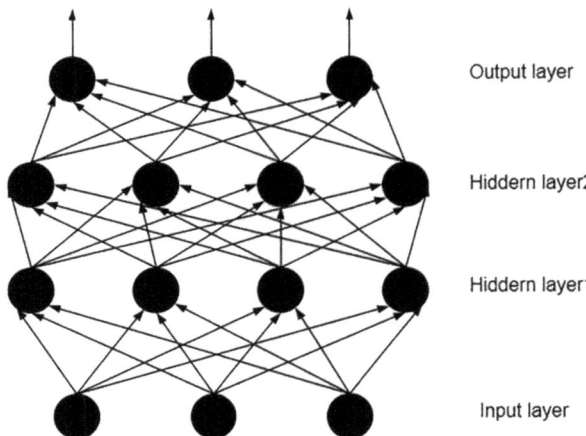

Fig. 4.16 Schematic diagram of the deep neural network structure

4.2 Deep Neural Network

Divide the training set. Because MATLAB support documents are for images or files, you can use the function to build a generator, data read into the toolbox directory under the helperModClassFrameGenerator, helperModClassFrameStore, etc. as a reference. These functions can be used to directly disrupt the data set, divided into training set, validation set and test set.

```
% main
xTrain = cell(12000,1);
yTrain = cell(12000,1);
for ii = 1:12000
xTrain{ii,1} = x(ii:ii+2000-1)
yTrain{ii,1} = y(ii)
end
```

Among them, xTrain and yTrain are two cellular arrays, which store the training data, xTrain is the network input, and yTrain is the output.

Build the network model. Build MATLAB neural network model layer construction and Keras is similar to the way, but less than Keras a lot of content, the new layer will be a little more difficult than Keras, MATLAB can also be read directly from the model built by Python, the specific operation of the relevant information can be consulted. The experimental network is a 3-layer fully connected network, which can be expressed as follows: input → fully connected → tanh → fully connected → output. The code is realized as follows.

```
function net=k_models(inputsize)
    net=[
sequenceInputLayer(inputsize,'Normalization','none','Name','Input Layer')
        fullyConnectedLayer(50,'Name','fc1')
        tanhLayer('Name','tanh1')
        fullyConnectedLayer(1,'Name','fc2')
        regressionLayer('Name','Output Layer')
    ];
end
```

In this case, the final output layer defines the loss function and the overall network can also be generated using the MATLAB deep network designer toolbox. MATLAB provides a function to visualize the network.

```
% main
modnet = k_models(2000)
analyzeNetwork(modnet)
```

Configure the learning process:

```
% main
options=trainingOptions('sgdm','InitialLearnRate',lr,'MaxEpoch
s',50,'MiniBatchSize',32,
   'Shuffle','every-epoch','Plots','training-progress','Verbose',0,
'LearnRateSchedule','piecewise',
   'LearnRateDropFactor',0.1,'LearnRateDropPeriod',0.1,'Execution
Environment','gpu')
```

The function trainingOptions() contains all the training information, such as gradient descent, learning rate, Epoch, BatchSize, etc., where Epoch is the whole data by default.

Verbose means whether to print the training process in the command line, turn off the program can improve the running time; Plots means whether to visualize the training effect, you can open the view; ExecutionEnvironment is the running environment using the CPU or GPU. other parameters can be found in the trainingOptions() function.

Training the model is very simple, given the input, output, model, training information can be.

```
% main
model = trainNetwork(xTrain,yTrain,modnet,options)
```

Test model. xTest is the test set, which is also a tuple array; the outputs z and yTrain are in the same format, the same tuple array, and there should not be any overlap between the test set and the training set, otherwise it will affect the final test accuracy.

```
% main
z = predict(model,xTest)
```

At this point, the basic process of building a deep neural network model has been completed, and we can master how to set the network parameters in a better and more reasonable way by referring to more examples.

AlexNet

AlexNet is a large-scale deep neural network created by Alex Krizhevsky, Ilya Sutskever and Geoffrey Hinton, which won the 2010 and 2012ILSVRC(ImageNet Large-scale Visual Recognition Challenge). 2012 was the first year that deep neural networks achieved a Top-5 error rate of 15.4% (the probability that a given image's label is not among the five outcomes the model considers most likely), and the second-place error rate was 26.2%. AlexNet was also a turning point in the resurgence of deep learning and neural networks, and it was because AlexNet won the ImageNet competition that deep learning officially entered the field of academia.

4.2 Deep Neural Network

AlexNet uses two NVIDIA GTX 580 GPUs for training. A single NVIDIA GTX 580 GPU has only 3 GB of memory, which limits the maximum size of the network that can be trained on. There are 1.2 million training samples in LSVRC2010 dataset. This training sample is too large to be placed on one GPU for training. Therefore, the AlexNet network is generally distributed on two GPUs, because the NVIDA GTX 580 GPUs can read and write to each other's memory directly, without the need to parallelize through host memory. Therefore, it is particularly suitable for cross GPU parallelization.

During AlexNet training, ReLU function is generally used as the activation function. The ReLU function does not need to normalize the input to prevent saturation. As long as the training sample generates a positive input to the ReLU function, the neuron will start learning. However, in the paper on AlexNet, it is proposed to use Local Response Normalization (LRN) to improve the input of neurons. After the pixel with coordinates (x, y) is convolved in the ith kernel function, the neuron activated by the nonlinearity of ReLU function is recorded as $a^i_{x,y}$, and $b^i_{x,y}$ is the neuron normalized by the local response corresponding to $a^i_{x,y}$. $b^i_{x,y}$ is calculated as follows:

$$b^i_{x,y} = a^i_{x,y} / \left[k + \alpha \sum_{j=\max\left(0, i-\frac{n}{2}\right)}^{\min\left(N-1, i+\frac{n}{2}\right)} \left(a^i_{x,y}\right)^2 \right]^\beta \qquad (4.21)$$

Where, n represents the number of nearest neighbor kernel maps on the same coordinate with $a^i_{x,y}$, and N is the depth of the kernel function of this layer. K. α and β are super parameters.

Since the order of the kernel mapping is arbitrary and has been determined before training, this local response normalization realizes horizontal suppression, resulting in competition between the outputs of neurons calculated using different kernels. The value with larger original output becomes larger, while the value with smaller original output becomes smaller, that is, suppression is achieved. Under this mechanism, the model generalization ability is improved.

In AlexNet, this technology is used in convolution layer 1 and convolution layer 2, where the parameter settings are all the same:

$$k = 2, n = 5, \alpha = 10^{-4}, \beta = 0.75$$

Figure 4.17 shows the AlexNet architecture. It can be clearly seen that this is an 8-layer neural network. The AlexNet architecture is introduced layer by layer below.

The input dimension of Conv1 is $224 \times 224 \times 3$, and the starting point of AlexNet's input is convolution layer 1(Conv1). The kernel dimension is $11 \times 11 \times 3 \times 96$, indicating that a 3-channel filter with a size of 11 is used for convolution with the input original image. The convolution kernel moves along the two axes of the original image, with a stride of 4 pixels. This layer does not use padding.

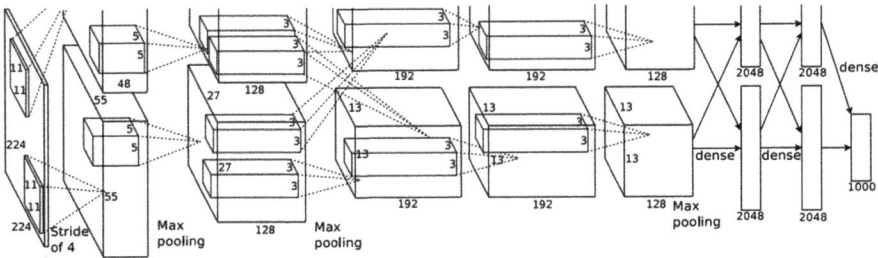

Fig. 4.17 AlexNet architecture

According to the AlexNet network architecture shown in Fig. 4.17, the dimension after convolution is 55 × 55 × 96, so the dimension of the input image should be 227 × 227 × 3. It can be seen that the input image undergoes preprocessing after being input to AlexNet, expanding the original dimension from 224 × 224 × 3 to 227 × 227 × 3. The following mainly takes the processed image as input.

The image size after convolution is

$$\frac{227-11}{4}+1=55$$

The image dimension after convolution is 55 × 55 × 96. The 55 × 55 × 96 convolution results were randomly divided into two groups, namely two 55 × 55 × 48. Then, use ReLU nonlinear activation function on two GPUs to activate and send to the maximum pooling layer. The kernel dimension of the max pooling layer after Conv1 is 3 × 3, and the step size is 2, so the size of the input image is

$$\frac{55-3}{2}+1=27$$

The output consists of two images with dimensions of 27 × 27 × 48, which are merged to form an image with dimensions of 27 × 27 × 96. After applying LRN on the Conv1 layer, the max pooling operation is performed.

Convolutional layer 2 (Conv2) has an input dimension of 2 × 27 × 27 × 48 and a convolution kernel size of 2 × 27 × 27 × 96. The step size is 1, and full 0-filling was performed in this layer, with 2 pixels at the top, bottom, left, and right. The size of the image after the convolution operation is

$$\frac{27-5+2\times 2}{1}+1=27$$

That is, the image dimensions of the image input on the two GPUs after convolution are 27 × 27 × 128. The convolution result is activated by ReLU function and sent to the max pooling layer, where the kernel dimension is 3 × 3 and the step size is 2. The image size obtained after the max pooling layer on the 2 GPUs is

$$\frac{27-3}{2}+1=13$$

That is, the image dimension is 13 × 13 × 128. Like the Conv1 layer, the Conv2 layer also operates on the max pooling layer after the local response is normalized. The results of the two GPUs are merged, and the dimensions of the image are 13 × 13 × 256.

The input dimension of convolutional layer 3 (Conv3) is 13 × 13 × 256 and the size of the convolutional kernel is 3 × 3 × 256 × 384. The step size is 1 and full 0-filling is performed in this layer with 1 pixel at the top, bottom, left and right. The size of the image after convolution operation is

$$\frac{13-3+2\times 1}{1}+1=13$$

The image dimension after convolution of the image input is 13 × 13 × 384. The convolved result is activated by ReLU function.

The input dimension of convolutional layer 4 (Conv4) is 13 × 13 × 384, and the size of the convolutional kernel is 3 × 3 × 384 × 384. The step size is 1, and full 0-padding is performed in this layer, with 1 pixel on the top, bottom, left, and right. The size of the image after the convolution operation is

$$\frac{13-3+2\times 1}{1}+1=13$$

The image input is convolved with a dimension of 13 × 13 × 384, and the convolved result is activated by the ReLU function, followed by local response normalization. The processed results are divided into two groups and sent to two GPUs, and the dimensions of the divided images are 13 × 13 × 192.

Convolutional layer 5 (Conv5) has an input dimension of 2 × 13 × 13 × 192 and a convolutional kernel size of 2 × 3 × 3 × 192 × 128. The step size is 1, and full 0-filling was performed in this layer, with 1 pixel at the top, bottom, left, and right. The size of the image after the convolution operation is

$$\frac{13-3+2\times 1}{1}+1=13$$

The dimension of the image input after convolution is 13 × 13 × 128. The convolution result is activated by ReLU function and goes to the maximum pooling layer operation with kernel dimension of 3 × 3 and step size of 2. After maximum pooling the image dimension is

$$\frac{13-3}{2}+1=6$$

That is, the image dimension of the layer was 2 6 × 6 × 128 and local response normalization was performed. By combining the 2 results, the image dimension becomes 6 × 6 × 256.

The network of fully connected layer 1 (FC1) can be regarded as a simple feedforward neural network with an input dimension of 6 × 6 × 256. In order to perform forward propagation, the input image matrix must be pulled into a one-dimensional vector, i.e., transformed into 1 × (6 × 6 × 256) = 1 × 9216. The dimension of the weights between the Conv5 layer and the FC1 layer is 9216 × 4096, and the dimension of the bias is 1 × 4096. The dimension of the forward propagation result is 1 × 4096, and the activation function is ReLU function. At the same time, the weights between FC1 and FC2 layers are Dropout processed with probability 0.5.

The input dimension of the fully connected layer 2 (FC2) is 1 × 4096, the dimension of the weights between FC1 and FC2 layers is 4096 × 4096, and the dimension of the bias is 1 × 4096. The dimension of the forward propagation result is 1 × 4096. The activation function is the ReLU function. The Dropout layer is also added to the process, with a probability of 0.5.

The input dimension of fully connected layer 3 (FC3) is 1 × 4096, the dimension of the weight between FC2 and FC3 is 4096 × 1000, the dimension of the bias is 1 × 1000, and the dimension of the forward propagation result is 1 × 1000. The activation function is a Softmax function. The 1000 dimensional vector represents the corresponding probability of classification in 1000.

Dropout temporarily "discards" the output of each hidden neuron from the network with a certain probability. Neurons that are "kicked out" in this way will not participate in forward propagation, nor will they join backward propagation. Thus, each time an input is given, the neural network adopts a different structure, but all these structures share the same weights. This technique reduces the complexity of joint adaptation between neurons because one neuron does not depend on the existence of another specific neuron and is more robust in connecting multiple different random subsets of other neurons.

Example 4.5 Use AlexNet to train and test the Kaggle dog and cat dataset.

Solution As shown in Fig. 4.18, Kaggle dog and cat data set for training images for 12,500, each image size is different, the image naming format for the "label + label". The two types of images were placed in two folders, and the folders were named with labels, so as to facilitate the use of MATLAB's own function to construct the dataset. The software used is MATLAB. MATLAB has provided an implementation of the AlexNet framework, which needs to be downloaded in advance.

1. Read the original dataset:

    imds = imageDatastore('E:\kaggle\train', ...

    'IncludeSubfolders',true, ...

    'LabelSource','foldernames');
    In the imds=imageDatastore(location,Name,Value) function, location refers to the location of the data set (folder address);IncludeSubfolders indicates that the

4.2 Deep Neural Network

Fig. 4.18 Partial images of Kaggle cat and dog dataset

subfolder contains the tag (using the folder classification and naming mentioned above) true is specified,false is not specified, default is not specified;FileExtensions is an image file extension that specifies that a particular type of image file should be read.

2. Read the number of datasets:
numTrainImages = numel(imds.Labels);
3. AlexNet's default input image size is $227 \times 227 \times 3$, where 3 refers to the three channels of the color image. So it is necessary to standardize the scale and pixel size of the images in the dataset:
for i = 1:numTrainImages
s = string(imds.Files(i));
I = imread(s);
I = imresize(I,[227,227]);
imwrite(I,s);
s
end
4. The original dataset imds is partitioned into training dataset imdsTrain and test dataset imdsValidation according to 7:3, and the partitioning is randomly proportional to:
[imdsTrain,imdsValidation] = splitEachLabel(imds,0.7,'randomized');
5. Read the original AlexNet:
net = alexnet;
inputSize = net.Layers(1).InputSize
layersTransfer = net.Layers(1:end-3);
numClasses = numel(categories(imdsTrain.Labels));

inputSize is the size of the input image for reading this network, layersTransfer indicates the acquisition of the network beyond the last three layers of AlexNet, which is kept unchanged, numClasses indicates the acquisition of the number of labels of the identified dataset (i.e., the number of classifications, the number of dog and cat classifications is 2).

6. Change the final fully connected layer to form a new network:
   layers = [
   layersTransfer
   fullyConnectedLayer(numClasses,'WeightLearnRateFactor',20,'BiasLearnRate Factor',20)
   softmaxLayer
   classificationLayer];

7. Train the AlexNet network:
   options = trainingOptions('sgdm', ...
   'MiniBatchSize',10, ...
   'MaxEpochs',6, ...
   'InitialLearnRate',1e-4, ...
   'Shuffle','every-epoch', ...
   'ValidationData',augimdsValidation, ...
   'ValidationFrequency',3, ...
   'Verbose',false, ...
   'Plots','training-progress');

8. Specify training options. For migration learning, it is necessary to retain the features in the shallower layers of the pre-trained network (migrated layer weights). To slow down the learning in the migrated layers, set the initial learning rate to a smaller value. Increasing the learning rate factor of the fully connected layer in step 7 can speed up the learning in the new final layer. This combination of learning rate settings will only speed up the learning in the new layer and slow down the learning for the other layers. The number of training rounds required to perform transfer learning is relatively small. A training round is a complete training cycle for the entire training dataset. The software validates the network every ValidationFrequency iteration during training.

9. Start training:
   netTransfer = trainNetwork(augimdsTrain,layers,options);
   As shown in Fig. 4.19, due to the structure of AlexNet itself, the accuracy fluctuates around 92%, and transfer learning can save a lot of time, so it is possible to use a data set with a small sample.

10. Randomly select 20 pieces from the mentioned test dataset as the dataset idx for this test and test network (Fig. 4.20):

    idx = randperm(numel(imdsValidation.Files),20);

    [YPred,scores] = classify(netTransfer,augimdsValidation);

    figure

    for i = 1:20

    subplot(5,4,i)

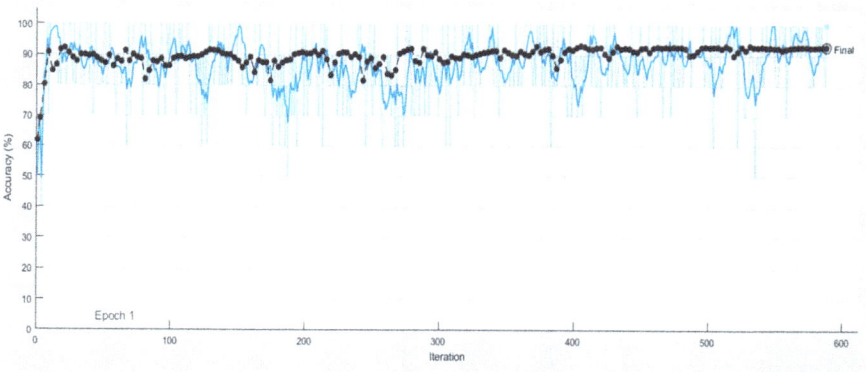

Fig. 4.19 Variation of accuracy during training

```
I = readimage(imdsValidation,idx(i));
imshow(I)
label = YPred(idx(i));
title(string(label));
end
```

4.2.3 Characteristics of Deep Neural Networks

Compared with the perceptual machine, DNN has added a hidden layer, which can have multiple layers to enhance the expressive ability of the model and also increase the complexity of the model. The neurons in the output layer can have multiple outputs, so the model can be flexibly applied to classification and regression. The number of layers can be increased appropriately to increase the accuracy, but increasing the number of layers too much may be counterproductive and bring problems such as the number of parameter expansion, etc. DNN neural networks are also applicable to other fields such as downscaling and clustering, and the expressive ability of neural networks can be further strengthened through the use of different activation functions compared with perceptual machines. In addition to these advantages, DNNs have the following limitations.

The number of parameters expands: Since DNN adopts a fully connected form, the connections in the structure bring a larger order of magnitude weight parameters, which not only easily leads to overfitting, but also easily falls into local optimization.

Local optimization: With the deepening of the neural network, the optimization function is more likely to fall into local optimization and deviate from the true global optimization. For limited training data, the performance is even worse than that of the shallow network.

Fig. 4.20 Some of the results obtained from the tests

Gradient explosion: Because the initial weight is too large, the front layer will change faster than the back layer, which will lead to more and more weight, and the phenomenon of gradient explosion occurs. In deep networks or recurrent neural networks, error gradients can accumulate in updates, turning into very large gradients, which then lead to large updates in the network weights and thus make the network unstable. In extreme cases, the value of the weight becomes so large that it overflows.

Gradient disappearance: The gradient of the front layer changes less than that of the back layer, so it changes more slowly, causing the gradient disappearance problem. Because the activation function used by the neural network is the Sigmoid function, the Sigmoid function can map negative infinity to positive infinity to [0,1], and the derivative of the function is $f'(x) = f(x)(1 - f(x))$. So when two numbers [0,1] are multiplied together, the result becomes small. The backpropagation of the neural network is to multiply the partial derivative of the function layer by layer. Therefore, when the number of layers of the neural network is very deep, the deviation generated by the last layer will become smaller and smaller because of the multiplication of many numbers less than 1, and eventually will become 0, resulting in the weight of the relatively shallow layer is not updated, which is the disappearance of the gradient.

4.2.4 Applications of Deep Neural Networks

Currently, DNNs have been widely used in various fields, and the following is a list of fields in which DNNs have already had a far-reaching impact and those in which they are likely to have a great impact in the future.

Images and Video: Video is probably the most prolific resource in the age of big data. It accounts for 70% of today's Internet traffic. For example, eight billion hours of surveillance video are generated worldwide every day. Computer vision needs to extract meaningful information from video. DNNs have greatly improved the accuracy of many computer vision tasks, such as image classification, object localization and detection, image segmentation, and action recognition.

Speech and language: DNNs have also dramatically improved the accuracy of speech recognition and many other related tasks, such as machine translation, natural language processing, and audio generation.

Medicine: DNN has played an important role in genetics, exploring the genetic aspects of many diseases such as autism, cancer and spinal muscular atrophy. It is also used in medical imaging to detect skin, brain and breast cancers.

Games: Recently, many difficult AI challenges, including games, have been solved using DNN methods. These successes require innovations in training techniques as well as reinforcement learning (where the network is trained with feedback from its own outputs). DNNs have already surpassed human accuracy in games such as Atari and Go.

Robotics: DNN has also been successful in a number of robotics tasks, such as robotic arm grasping, motion planning, visual navigation, stability control of quadcopters, and driving strategies for unmanned vehicles. DNN has a wide range of applications today. Looking to the future, DNNs will play an even more important role in medicine and robotics. There will also be more applications in finance (e.g., trading, energy forecasting and risk assessment), infrastructure development (e.g., structural safety, traffic control), weather forecasting and event detection.

Embedded vs. Cloud: Different DNN applications and processes (training vs. inference) have different computational requirements. In particular, the training process requires a large dataset and a lot of computational resources for iteration, so it needs to be computed in the cloud. The inference process can be performed in the cloud or at the endpoints (e.g., IoT devices or mobile terminals). In many applications of DNNs, where the inference process needs to be done in the vicinity of the sensors, such as driverless cars, drone navigation, or robots, the processing has to be done locally because of the high security risk due to latency and transmission instability. The computational complexity of processing video for DNN applications is considerable, so low-cost hardware that can efficiently analyze video remains an important constraint for DNN applications. Embedded platforms capable of performing DNN inference processes are subject to stringent energy consumption, computation and storage cost constraints.

4.2.5 Optimization of Deep Neural Networks

The deeper the DNN neural network, the better the performance is not necessarily. A new activation function can be used and the learning rate can be adjusted when the training data is not performing well.

Because of the limitation of Sigmoid function, the output value is in the range of 0 ~ 1, and it changes slowly in the place near 0 and 1. When the neural network has more layers, the gradient of the sigmoid function changes slowly after many back propagations, and the learning efficiency is very low. In this case, a new activation function such as ReLU or maxout can be used as the transfer function.

The ReLU activation function image is shown in Fig. 4.21. The ReLU activation function is uniformly transformed, which can effectively mitigate the gradient vanishing problem. The maxout function is a learnable activation function, which can be any segmented linear convex function. Maxout activation function needs to be composed of two or more nodes to form a group, and outputs the maximum value in the output process.

For learning rate adjustment the RMSProp method can be used, and the weight update formula is as follows:

$$\begin{cases} w^1 \leftarrow w^0 - \dfrac{\eta}{\sigma^0} g^0 & \sigma^0 = g^0 \\ w^2 \leftarrow w^1 - \dfrac{\eta}{\sigma^1} g^1 & \sigma^1 = \sqrt{\alpha(\sigma^0)^2 + (1-\alpha)(g^1)^2} \\ w^3 \leftarrow w^2 - \dfrac{\eta}{\sigma^2} g^2 & \sigma^2 = \sqrt{\alpha(\sigma^1)^2 + (1-\alpha)(g^2)^2} \\ \cdots \\ w^{t+1} \leftarrow w^t - \dfrac{\eta}{\sigma^t} g^t & \sigma^t = \sqrt{\alpha(\sigma^{t-1})^2 + (1-\alpha)(g^t)^2} \end{cases} \quad (4.22)$$

Fig. 4.21 ReLU activation function image

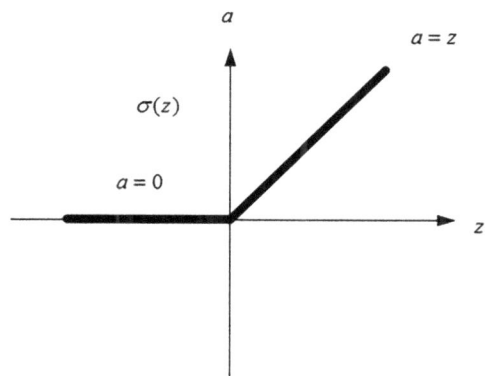

By adjusting α so that whether the next weight adjustment is affected more by the inverse of the gradient or more by the previous adjustment. The value of α can also be learned by the gradient descent method.

4.3 Convolutional Neural Networks

Convolutional Neural Networks (CNNs) originated from the study of the visual cortex of the brain and have been used for image recognition since the 1980s. In the past few years, due to the increase in computer computational power, the increase in the amount of available training data, and the increase in the number of techniques used for deep network training, CNNs have achieved hyperhumanization in some complex visual tasks and have been widely used in image search services, self-driving cars, and automated video classification systems, among others. In addition, not limited to visual perception, CNNs have also been successfully used for other tasks such as speech recognition or Natural Language Processing (NLP).

The main topics covered in this section include: the history and basic concepts of convolutional neural networks; the structure of convolutional neural networks; application examples; and common convolutional neural networks.

4.3.1 History and Basic Concepts of Convolutional Neural Networks

CNN is a kind of feedforward neural network with convolutional computation and deep structure, and it is one of the representative algorithms of deep learning. CNN has the ability of representation learning to classify input information according to its hierarchical structure, so it is also called shift-invariant artificial neural networks (SIANN).

CNN is an efficient recognition method that has been developed in recent years and attracted wide attention. In the 1960s, Hubel and Wiesel proposed CNN when they studied the neurons used for local sensitivity and direction selection in the cat cortex, and found that their unique network structure could effectively reduce the complexity of the feedback neural network. Nowadays, CNN has become one of the research hotspots in many scientific fields, especially in the field of pattern classification. Since this network avoids complex pre-processing of images and can be directly input to the original image, it has been more widely used. The new recognition machine proposed by K. Fukushima in 1980 was the first realization of convolutional neural network. Subsequently, more researchers have improved the network. A representative research result is the "Improved Cognitive Machine" proposed by Alexander and Taylor, which combines the advantages of various improved methods and avoids the time-consuming error back-propagation.

The basic structure of a CNN consists of two layers.

The feature extraction layer, where the input of each neuron is connected to the local receptive domain of the previous layer and that local feature is extracted. Once the local feature is extracted, its positional relationship with other features is also determined.

Feature mapping layer, each computational layer of the network consists of multiple feature mappings, each feature mapping is a plane, and all neurons on the plane have equal weights. The general feature mapping structure adopts the Sigmoid function, which has a small kernel of influence function, as the activation function of the convolutional network, so that the feature mapping has displacement invariance. In addition, the number of free parameters of the network is reduced because the neurons on a mapping surface share the weights. Each convolutional layer in the convolutional neural network is immediately followed by a computational layer for local averaging and quadratic extraction, and this unique two-feature extraction structure reduces the feature resolution.

CNN is mainly used to recognize two-dimensional graphs with displacement, scaling and other forms of distortion invariance, and this part of the function is mainly realized by the pooling layer. Since the feature detection layer of CNN learns from the training data, the use of CNN avoids explicit feature extraction and learns implicitly from the training data. Moreover, since the neurons in the same feature map have the same weight, the network can learn in parallel, which is also a major advantage of convolutional networks over networks with neurons connected to each other. Convolutional neural network has a unique superiority in speech recognition and image processing with its special structure of local weight sharing. Its layout is closer to the actual biological neural network, and the weight sharing reduces the complexity of the network. In particular, multi-dimensional input vectors can be directly input into the network, which avoids the complexity of data reconstruction in the process of feature extraction and classification.

Nowadays, convolutional neural networks continue to develop in many directions, and have made breakthroughs in speech recognition, face recognition, general object recognition, motion analysis, natural language processing, and even brainwave analysis.

4.3.2 Structure of Convolutional Neural Network

A CNN is very similar to a regular neural network, consisting of neurons where weights can be learned from the data, and each neuron receives some input and performs a dot product calculation. The last fully connected layer has a loss function on it, and nonlinear functions can be used. Conventional neural networks receive input data as a single vector passed to a series of hidden layers. Each hidden layer contains a group of neurons, each of which is connected to all the other neurons in the previous layer. In a single layer, each neuron is completely independent and does not share any connections. The last fully connected layer, the output layer, contains the classification score in the image classification problem. In general, a classical CNN network structure contains an input layer, a convolutional layer, a pooling layer, a fully connected layer, and an output layer (Zelinsky 2015). A conventional 6-layer convolutional neural network is shown in Fig. 4.22.

Input Layer

The input layer receives the image data. If the classical neural network model is used, the whole image needs to be read as input to the neural network model (i.e., the fully connected approach). The larger the size of the image, the more parameters are connected to it, resulting in a very large amount of computation.

Human cognition of the outside world is generally from the local to the global, first to the local perceptual understanding, and then gradually to the global cognition, which is the human cognition mode. The spatial connection in the image is similar, pixels in the local range are more closely related to each other, while pixels in the distance are less relevant. Therefore, it is not necessary for each neuron to use the global receptive field shown in Fig. 4.23a to perceive the global image, but only need to perceive the local information, and then synthesize the local information at a higher level to obtain the global information. This pattern is an important module in a convolutional neural network that reduces the number of parameters: the local receptive field, as shown in Fig. 4.23b.

Fig. 4.22 6-layer convolutional neural network

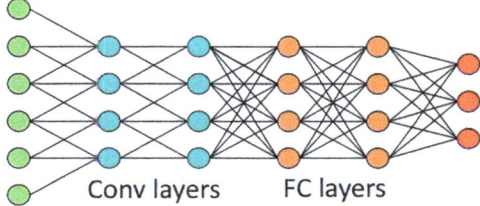

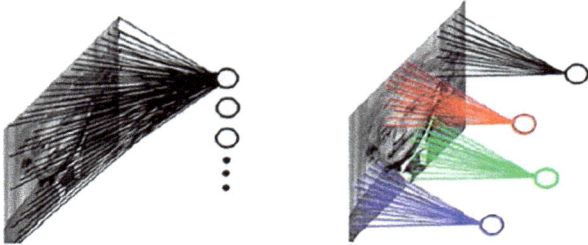

Fig. 4.23 Receptive field

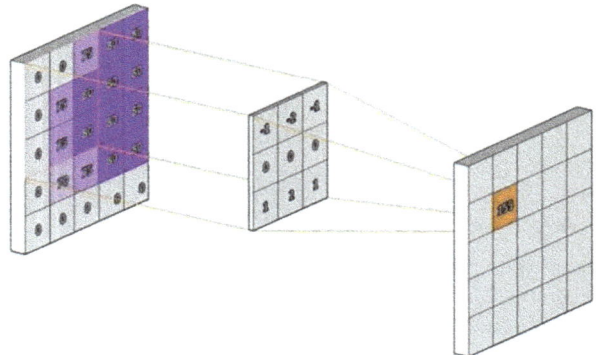

Fig. 4.24 Convolution operation

Convolutional Layer

As the input information, the matrix of image is often very large, and if the fully connected neural network is used for training, the calculation amount is very large. Therefore, CNN is proposed, and one of its highlights is the convolutional layer. The main purpose of the convolutional layer is to extract features from the input image.

When given a new image, CNN does not know exactly what parts of the original image these features will match, so it will try every possible position in the original image, equivalent to turning the feature into a filter. This matching process is called a convolution operation and is shown in Fig. 4.24.

Through the convolution operation of each feature, a new two-dimensional array is obtained, called the feature map. The closer the value is to 1, the more complete the matching of the corresponding position and feature is; the closer the value is to -1, the more complete the reverse matching of the corresponding position and feature is; and the closer the value is to 0, the more complete the matching of the corresponding position and feature is.

4.3 Convolutional Neural Networks

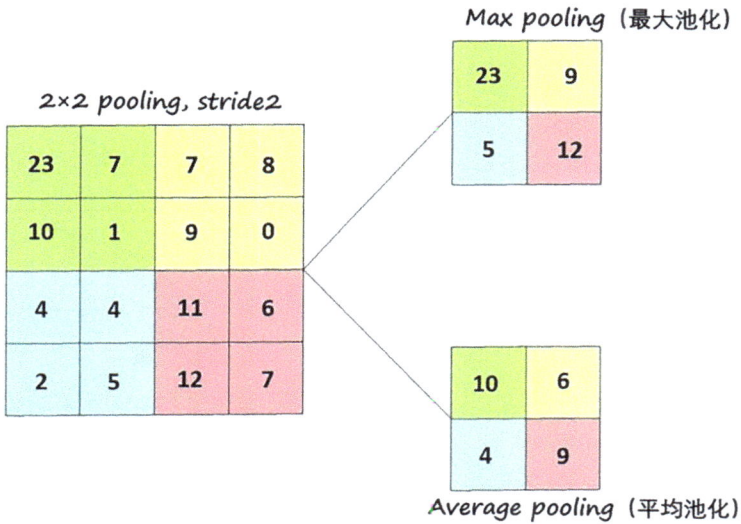

Fig. 4.25 Pooling operation

Pooling Layer

Once you know how the convolutional layer works, the pooling layer is easy to understand. Their purpose is to reduce the computational load, memory utilization, and number of parameters (thereby reducing the risk of overfitting) by sampling the input image twice. Reducing the size of the input image also allows the neural network to tolerate a certain image shift (position invariance).

Generally, the size of the pooling area is 2 × 2, and then it is converted into the corresponding value according to certain rules, such as the max-pooling value and mean-pooling value of the pooling area, and this value is taken as the pixel value of the result, as shown in Fig. 4.25.

Fully Connected Layer

The fully connected layer plays the role of "classifier" in the whole convolutional neural network. If operations such as convolution layer, pooling layer, and activation function layer map the original data to the hidden layer feature space, the full join layer maps the learned feature representation to the sample's label space. That is, after deep network such as convolution, activation function and pooling, the results are identified and classified through the full connection layer.

First, string the results of the deep network after convolution, activation function, and pooling, as shown in Fig. 4.26.

Fig. 4.26 Fully connected operation

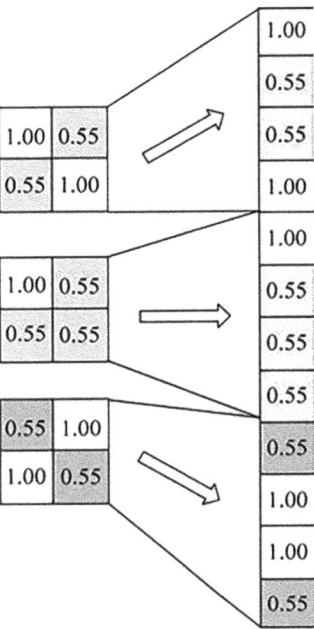

Next, the binary classification task is carried out. According to the weights obtained by the model training mentioned above and the results calculated by deep network such as convolution, activation function and pooling, weighted summation is carried out. After that, the predicted values of each result are obtained through Softmax function. Then the largest value is taken as the result of recognition (as shown in Fig. 4.27, the recognition value of letter X is finally calculated as 0.92, and the recognition value of letter O is 0.51, then the result is determined as letter X).

The output layer is the output node of the prediction classification, each node represents a classification, the letters X and O represent the model of two classifications, and the incentive function of each node is

$$\sigma_i(z) = \frac{e^{z_i}}{\sum_{j=1}^{m} e^{z_j}} \quad (4.22)$$

Where, i is the subscript order of the output node; m is the number of output nodes; z_i is

$$z_i = w_i x + b \quad (4.23)$$

4.3 Convolutional Neural Networks

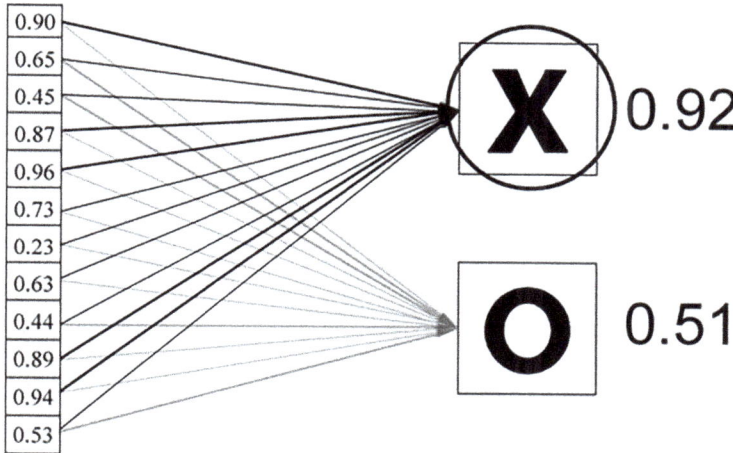

Fig. 4.27 Fully connected layer and output layer

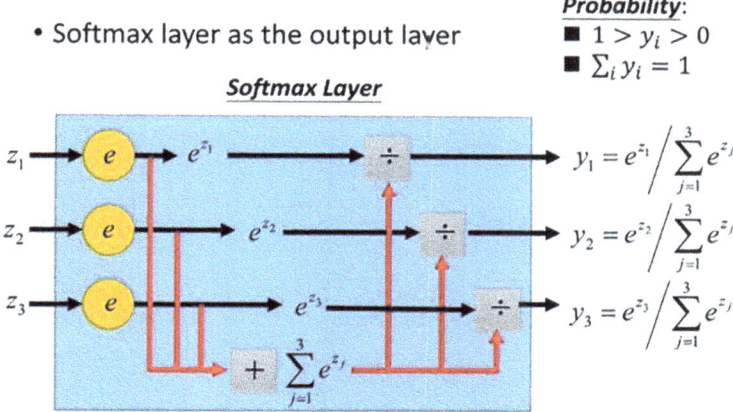

Fig. 4.28 Softmax layer

Among them,

$$\sum_{i=1}^{j}\sigma_i(z) = 1 \qquad (4.24)$$

The output of the full connection layer is mapped to the probability value through Softmax function, and the node with the highest probability is selected as the prediction target. The calculation process is shown in Fig. 4.28.

4.3.3 Application of Convolutional Neural Network and MATLAB Example

Example 4.6 Use the handwritten digit set in the MINIST database to train the CNN to correctly recognize the digits in it.

Solution Train with LeNet, a lightweight CNN that contains the basic building blocks of deep learning: a convolutional layer, a pooling layer, and a fully connected layer. Starting with a set of images for each of the 10 numbers, 700 samples are selected for each group to train the CNN, and then 100 samples for each group are tested to see how the CNN performs.

The MATLAB program is implemented as follows:

```
%% Program description
% 1 pooling pooling uses an average of 2×2
% 2 Description of the number of network nodes
% Input layer 28 x 28
% First layer 24×24 convolution ×6
% Second layer 12×12 pooling ×6
% Third layer 8×8 convolution ×16
% Fourth layer 4×4 pooling ×16
% Fifth layer is fully connected 40
% Sixth layer is fully connected 10
clear all;clc;
%% Network initialization
layer_c1_num=6;
layer_c2_num=16;
yita=0.05; % Weight adjustment step
bias=1;
% Convolution kernel initialization
[kernel_c1,kernel_c2]=init_kernel(layer_c1_num,layer_c2_num);
pooling_a=ones(2,2)/4; % Pooling layer kernel initialization
weight_full_1=rand(16,40)/sqrt(40); % Full connected
layer weights
weight_full_2=rand(40,10)/sqrt(10);
weight_c2=rand(6,16)/10;
weight_arr2num=rand(4,4,layer_c2_num)/sqrt(16);
disp('Network initialization complete......');
%% Start network training
disp('Start network training......');
for n=1:500
    for m=0:9
        % Read samples
```

4.3 Convolutional Neural Networks

```
            train_data=imread(strcat(num2str(m),'_',num2st
r(n),'.bmp'));
            train_data=double(train_data);
            % Normalization
            %train_data=train_data/sqrt(sum(sum(train_data.^2)));
            % Label settings
            label_temp=-ones(1,10);
            label_temp(1,m+1)=1;
            label=label_temp;
            for iter=1:10
                % Forward pass, enter convolutional layer1
                for k=1:layer_c1_num
state_c1(:,:,k)=convolution(train_data,kernel_c1(:,:,k));
                    % Enter pooling layer1
state_s1(:,:,k)=pooling(state_c1(:,:,k),pooling_a);
                end
                % Enter convolutional layer2
[state_c2,state_c2_temp]=convolution_c2(state_s1,kernel_
c2,weight_c2);
                % Enter pooling layer 2
                for k=1:layer_c2_num
state_s2_temp1(:,:,k)=pooling(state_c2(:,:,k),pooling_a);
                end
                % Turn matrices into numbers
                for k=1:layer_c2_num
state_s2_temp2(1,k)=sum(sum(state_s2_temp1(:,:,k).*weight_
arr2num(:,:,k)))+bias;
                    state_s2(1,k)=1/(1+exp(-state_s2_temp2(1,k)));
state_s2(1,k)=sum(sum(state_s2_temp1(:,:,k).*weight_
arr2num(:,:,k)));
                end
                % 16 feature numbers, enter the fully
connected layer1
                state_f1=state_s2*weight_full_1;
                % Enter the fully connected layer2
                state_f2=state_f1*weight_full_2;
                %% Error calculation part
                Error=state_f2-label;
                Error_Cost=sum(Error.^2);
                if(Error_Cost<1e-4)
                    break;
```

```
                end
            %% Parameter adjustment part
[kernel_c1,kernel_c2,weight_c2,weight_full_1,weight_
full_2,weight_arr2num]=CNN_upweight1(Error,train_data,...
                state_c1,state_s1,...
                state_c2,state_s2_temp1,...
                state_s2,state_s2_temp2,...
                state_f1,state_f2,...
                kernel_c1,kernel_c2,...
                weight_c2,weight_full_1,...
                weight_full_2,weight_arr2num,yita,state_c2_temp);
        end
    end
end
disp('The network is trained and the test begins......');
%% Test part
count_num=0;
for n=501:600
    for m=0:9
        % Read sample
        train_data_test=imread(strcat(num2str(m),'_',num2str(n),'.bmp'));
        train_data_test=double(train_data_test);
        % Forward pass, enter convolutional layer1
        for k=1:layer_c1_num
state_c1(:,:,k)=convolution(train_data,kernel_c1(:,:,k));
            % Enter pooling layer1
            state_s1(:,:,k)=pooling(state_c1(:,:,k),pooling_a);
        end
        % Enter convolutional layer2
[state_c2,state_c2_temp]=convolution_c2(state_s1,kernel_c2,weight_c2);
        % Enter pooling layer 2
        for k=1:layer_c2_num
state_s2_temp1(:,:,k)=pooling(state_c2(:,:,k),pooling_a);
        end
        % Turn matrices into numbers
        for k=1:layer_c2_num
state_s2_temp2(1,k)=sum(sum(state_s2_temp1(:,:,k).*weight_
```

4.3 Convolutional Neural Networks   113

```
arr2num(:,:,k)))+bias;
            state_s2(1,k)=1/(1+exp(-state_s2_temp2(1,k)));
state_s2(1,k)=sum(sum(state_s2_temp1(:,:,k).*weight_
arr2num(:,:,k)));
        end
        % 16 feature numbers, enter the fully connected layer1
        state_f1=state_s2*weight_full_1;
        % Enter the fully connected layer2
        state_f2=state_f1*weight_full_2
        [~,train_label]=max(state_f2);
        if(train_label-1==m)
            count_num=count_num+1;
            train_label
        end
    end
end
ture_rate=1.0*count_num/300;
fprintf('The recognition accuracy rate of this neural network for
MNIST sample database is    %4d%%    \n',ture_rate);
```

4.3.4 Latest Development of Convolutional Neural Network

The classic structure of CNN includes LeNet, AlexNet, ZFNet, VGG, NIN, GoogLeNet, ResNet and SENet. CNN has been widely used in the field of computer vision, and has achieved good results. Through the performance of CNN in ImageNet competition in recent years, it can be seen that in order to pursue classification accuracy, the model depth is getting deeper and deeper, and the model complexity is getting higher and higher, such as the number of layers of deep residual network (ResNet) has reached 152.

However, in some real-world application scenarios, such as mobile or embedded devices, such large and complex models are difficult to apply. First, the model is too large and faces the problem of insufficient memory. Second, these scenarios require low latency, or fast response times. Therefore, the study of small but efficient CNN models is crucial in these scenarios. The current research summary can be divided into two directions: one is to compress the trained complex model to get a small model; The second is to directly design small models and train them. In any case, the goal is to reduce the model size (parameters size) while increasing the model speed (low latency) while maintaining model performance. Google recently proposed a small but efficient CNN model that compromises accuracy and latency. The development process of CNN is shown in Fig. 4.29.

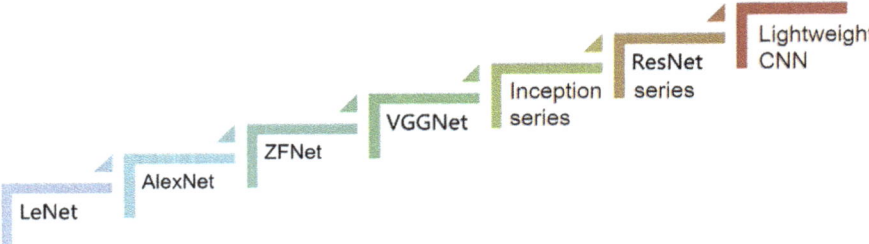

Fig. 4.29 Development of CNN

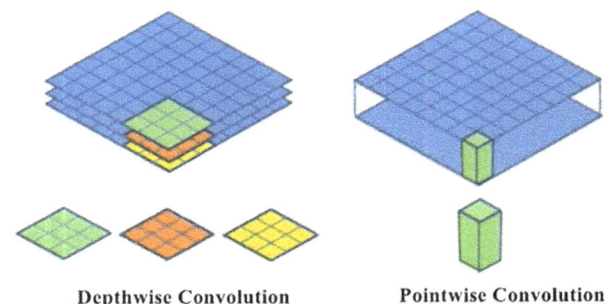

Fig. 4.30 Depthwise convolution and pointwise convolution

MobileNet

The core part of MobileNet is the depthwise separable convolution, which actually decomposes the original convolution layer into two parts: the depthwise convolution and a 1 × 1 convolution, namely the pointwise convolution, as shown in Fig. 4.30.

Depthwise convolution is different from standard convolution. The convolution kernel of standard convolution is applied to all input channels, while depthwise convolution adopts different convolution kernels for each input channel. That is to say, one convolution kernel corresponds to one input channel, so it is a depth operation. And pointwise convolution is an ordinary convolution, except that it uses 1 × 1 convolution kernels. For depthwise separable convolution, first, depthwise convolution is used to convolve different input channels separately, and then pointwise convolution is employed to combine the outputs above. The overall effect of this is similar to that of a standard convolution, but it can greatly reduce the amount of computation and the number of parameters in the model. The structures of various convolutions are shown in Fig. 4.31.

Table 4.2 presents the comparison among MobileNet, GoogLeNet and VGG16. Compared with VGG16, the accuracy of MobileNet drops by less than 1%, but it is superior to GoogLeNet. However, MobileNet has an absolute advantage in terms of the amount of computation and the number of parameters.

4.3 Convolutional Neural Networks

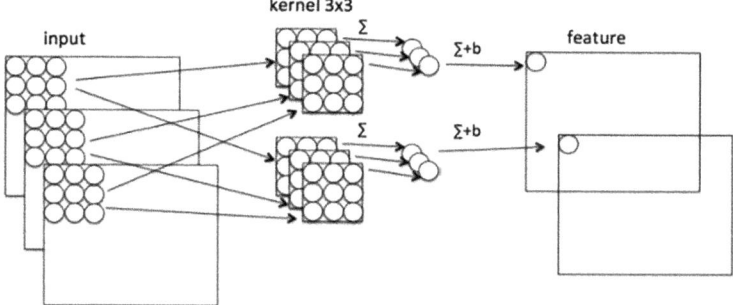

(a) Depthwise convolution

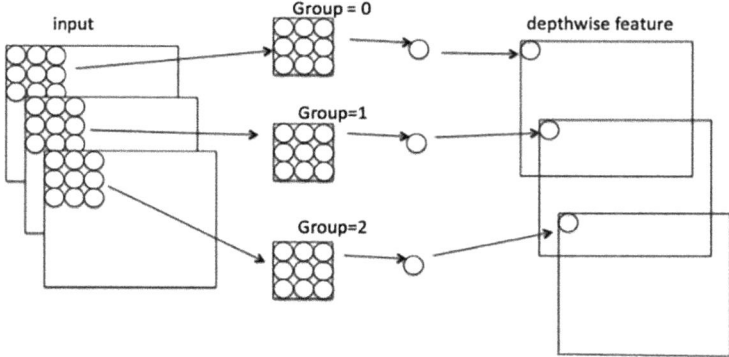

(a) Pointwise convolution

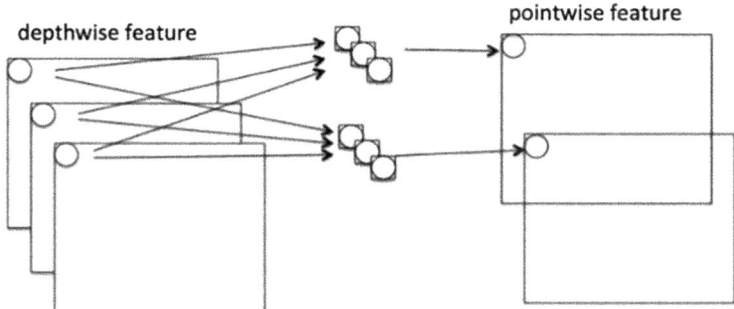

(a) Depthwise separable convolution

Fig. 4.31 Comparison of convolutional structures. (**a**) Depthwise convolution, (**b**) Pointwise convolution. (**c**) Depthwise separable convolution

Table 4.2 Performance comparison of MobileNet with GoogLeNet and VGG16

Model	ImageNet accuracy/%	Mult-Adds/million	Parameters/million
MobileNet-224	70.6	559	4.2
GoogleNet	69.8	1550	6.8
VGG 16	71.5	15,300	138

Attention Mechanism of CNN and Visual Model

The attention mechanism in deep learning is derived from the attention mechanism of the human brain (see Fig. 4.32). When the human brain receives external information (such as visual information and auditory information), it often does not process and understand all the information, but only focuses on some significant or interesting information, which is conducive to filtering out unimportant information and improving information processing efficiency.

The mechanism of attention is essentially similar to how humans observe things in the outside world. Generally speaking, when people observe external things, they tend to focus on some important local information of the observed things first, and then combine information from different areas to form an overall impression of the observed things. The attention mechanism can make deep learning more targeted when observing targets, and improve the accuracy of target recognition and classification.

The attention mechanism can help the model assign different weights to each part of the input, extract more critical and important information, and make the model make more accurate judgments, while not bringing more overhead to the calculation and storage of the model.

There are two types of attention mechanisms: soft attention and hard attention. The differences between soft attention and hard attention are as follows.

Soft attention pays more attention to regions or channels. Moreover, soft attention is deterministic attention, which can be directly generated by the network after learning. The most crucial point is that soft attention is differentiable. Differentiable attention enables the calculation of gradients through neural networks and the learning of attention weights via forward propagation and backward feedback. In the field of computer vision, relevant work in many areas (such as classification, detection, segmentation, generative models, video processing, etc.) is using soft attention. Typical representatives include SENet, SKNet and so on.

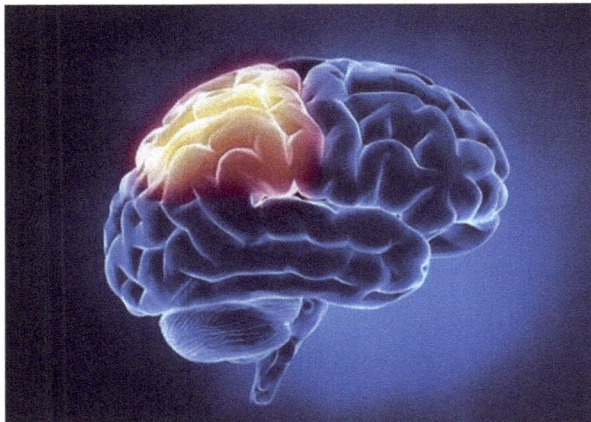

Fig. 4.32 Attention mechanism of human brain

4.3 Convolutional Neural Networks

Hard attention pays more attention to the points of an image. That is to say, each point in the image may give rise to attention. Meanwhile, hard attention is a stochastic prediction process and emphasizes dynamic changes more. Of course, the most crucial point is that hard attention is non-differentiable attention, and the training process is often completed through reinforcement learning.

Most of the research work on the combination of deep learning and visual attention mechanisms focuses on using masks to form attention mechanisms. The principle of masks lies in marking out the key features in image data through another layer of new weights. Through learning and training, deep neural networks can learn the regions that need to be focused on in each new image, thus forming attention.

The basic idea of the attention mechanism in computer vision is to make the model learn to concentrate and focus its attention on important information while ignoring unimportant information.

The essence of the attention mechanism is to utilize relevant feature maps to learn the weight distribution, and then apply the learned weights to the original feature maps for weighted summation, as shown in Fig. 4.33. However, there are slight differences in the way of applying weights, which can be roughly summarized into the following four points.

1. For soft attention, the weighting retains all components and weights them all; for hard attention, some components are selected in the distribution according to a certain sampling strategy, which is generally completed by using reinforcement learning.
2. The weighting can act on the spatial scale to weight different spatial regions.
3. Weighting can be applied at the channel level to assign weights to the features of different channels.
4. Weighting can be applied to historical features at different time instants. Weights can be added in combination with the recurrent structure, such as in tasks related to machine translation or video processing.

Fig. 4.33 Attention mechanism in machine vision

In order to introduce the attention mechanism in computer vision more clearly, the model structure of the attention mechanism is usually divided into three attention domains: spatial domain, channel domain and mixed domain.

Spatial domain makes corresponding spatial transformations to the spatial domain information in the image, so that the key information can be extracted. A mask is generated and scored for the space, such as spatial attention module (SAM).

The channel domain is analogous to adding a weight to each signal on a channel that represents the relevance of that channel to the key information. The larger the weight, the higher the correlation. Masks are generated and scored for channels, such as SENet and channel attention module (CAM).

The hybrid domain is a fusion of the spatial and channel domains.

Based on the model structure, spatial attention, channel attention, and mixed attention of space and channel are commonly used in convolutional neural networks.

For a convolutional neural network, each layer of the CNN outputs a C × H × W feature map, where C is the channel, which also represents the number of convolutional kernels and the number of features, and H and W are the height and width of the original image after compression. Spatial attention is to learn a weight for all the channels in the two-dimensional plane for the H × W size feature map, then each pixel will learn a weight. It can be imagined as a pixel is a vector in C dimensions with depth C. In C dimensions, the weights are the same, but in the plane, the weights are not the same. The structure of the spatial attention model is shown in Fig. 4.34.

For each channel, different weights are learned in the channel dimension, and the same weights are learned in the plane dimension, so the attention based on the channel domain is usually to directly pool the information in a channel globally, while ignoring the local information in each channel. Spatial attention and channel attention can be understood as focusing on different areas of the image and focusing on different features of the image. Channel attention in the image classification of network structure, a typical SENet. The structure of the channel attention model is shown in Fig. 4.35.

Channel attention code based on Pytorch platform:

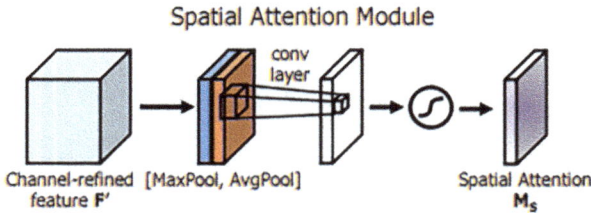

Fig. 4.34 Spatial attention module

4.3 Convolutional Neural Networks

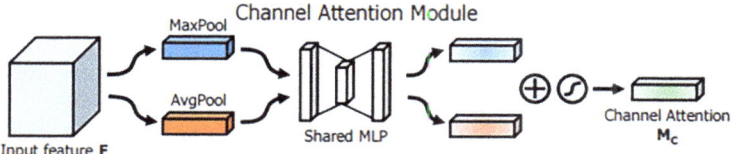

Fig. 4.35 Channel attention module

```
class SELayer(nn.Module):
    def __init__(self, channel, reduction=1):
        super(SELayer, self).__init__()
        self.avg_pool = nn.AdaptiveAvgPool2d(1)
        self.fc1 = nn.Sequential(
            nn.Linear(channel, channel // reduction),
            nn.ReLU(inplace=True),
            nn.Linear(channel // reduction, channel),
            nn.Sigmoid())
        self.fc2 = nn.Sequential(
            nn.Conv2d(channel , channel // reduction, 1, bias=False),
            nn.ReLU(inplace=True),
            nn.Conv2d(channel , channel // reduction, 1, bias=False),
            nn.Sigmoid()
        )
    def forward(self, x):

        b, c, _, _ = x.size()
        y = self.avg_pool(x).view(b, c)
        y = self.fc1(y).view(b, c, 1, 1)
        return x * y
```

The channel attention code based on Keras platform is as follows:

```
class SELayer():
    """
    SE layer contains Squeeze and excitaton operations
    """
    def __init__(self,input_tensor,ratio):
        """
        :param input_tensor: input_tensor.shape=[h,w,c]
        :param ratio:Number of output channels for excitation intermediate operation
        """
```

```python
        self.in_tensor=input_tensor
        self.in_channels=keras.backend.in_shape(input_tensor)[-1]
        self.ratio=ratio
    def squeeze(self, input):
        return GlobalAveragePooling2D()(input)
    def excitation_dense(self,input):

        out=Dense(units=self.in_channels//self.ratio)(input)
        out=Activation("relu")(out)
        out=Dense(units=self.in_channels)(out)
        out=Activation("sigmoid")(out)
        out=Reshape((1,1,self.in_channels))(out)
        return out
    def excitation_conv(self,input):

out=Conv2D(filters=self.in_channels//self.ratio,kernel_size=(1,1))
(input)
        out=Activation("relu")(out)
        out=Conv2D(filters=self.in_channels,kernel_
size=(1,1))(out)
        out=Activation('sigmoid')(out)
        out = Reshape((1, 1, self.in_channels))(out)
        return out
    def forward(self):

        """
        Use conv by default
        :param self:
        :return:
        """
        out=self.squeeze(self.in_tensor)
        out=self.excitation_conv(out)
        scale=multiply([self.in_tensor,out])
        return scale
#Perhaps
def se_layer(inputs_tensor=None,ratio=None,num=None,**kwargs):
    """
    SE-NET
    :param inputs_tensor:input_tensor.
shape=[batchsize,h,w,channels]
    :param ratio:
    :param num:
    :return:
    """
    channels = K.int_shape(inputs_tensor)[-1]
```

4.3 Convolutional Neural Networks

```
    x = KL.GlobalAveragePooling2D()(inputs_tensor)
    x = KL.Reshape((1, 1, channels))(x)
    x = KL.Conv2D(channels//ratio, (1, 1), strides=1, name="se_
conv1_" + str(num), padding="valid")(x)
    x = KL.Activation('relu', name='se_conv1_relu_'+str(num))(x)
    x = KL.Conv2D(channels, (1, 1), strides=1, name="se_
conv2_"+str(num), padding="valid")(x)
    x = KL.Activation('sigmoid', name='se_conv2_
relu_'+str(num))(x)
    output = KL.multiply([inputs_tensor, x])
    return output
```

The representatives of mixed Attention are BAM and CBAM (Convolutional Block Attention Module), where the structure of CBAM is shown in Fig. 4.36.

The input. First, through channel attention, an image passes through several convolution layers to obtain a feature matrix, the number of channels of this matrix is the number of convolution layer kernel, a common convolution kernel usually reaches 1024 or 2048. Not every channel is useful for information transfer, so the optimized characteristics are obtained by filtering (that is, paying attention to) these channels. The main idea is to increase the weight of effective channels and reduce the weight of invalid channels.

The global pooling operation is performed on the channel dimension, and the weights are obtained through the same MLP, which is added as the final attention vector (weights). CBAM is very similar to SENet, which has been proved to improve the effect in many papers. The difference here is that SENet adopts average pooling, while CBAM adds maximum pooling.

GSoP-Net

The overall structure of GSoP-Net is shown in Fig. 4.37.

After the input tensor is first reduced by convolution, the GSoP block calculates the covariance matrix, and then obtains the output tensor through two continuous operations of linear convolution and nonlinear activation. The output tensor scales the original input along the channel dimension, which is also a reflection of channel

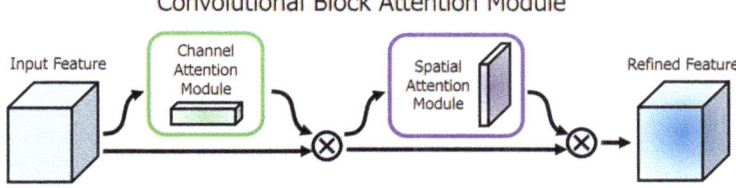

Fig. 4.36 CBAM structure

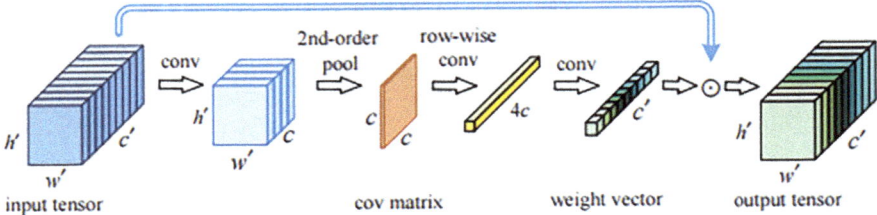

Fig. 4.37 Overall structure of GSoP-Net

attention to a certain extent. However, unlike SENet, GSoP proposes two-dimensional average pooling, which reflects the relationship between channels in the form of covariance.

The code implementation of GSoP-Net based on Pytorch is as follows:

```
class Covpool(Function):
    @staticmethod
    def forward(ctx, input):
        x = input
        batchSize = x.data.shape[0]
        dim = x.data.shape[1]
        h = x.data.shape[2]
        w = x.data.shape[3]
        M = h*w
        x = x.reshape(batchSize,dim,M)
        I_hat = (-1./M/M)*torch.ones(M,M,device = x.device) + (1./M)* torch.eye(M,M,device = x.device)
        I_hat = I_hat.view(1,M,M).repeat(batchSize,1,1).type(x.dtype)
        y = x.bmm(I_hat).bmm(x.transpose(1,2))
        ctx.save_for_backward(input,I_hat)
        return y
    @staticmethod
    def backward(ctx, grad_output):
        input,I_hat = ctx.saved_tensors
        x = input
        batchSize = x.data.shape[0]
        dim = x.data.shape[1]
        h = x.data.shape[2]
        w = x.data.shape[3]
        M = h*w
        x = x.reshape(batchSize,dim,M)
        grad_input = grad_output + grad_output.transpose(1,2)
        grad_input = grad_input.bmm(x).bmm(I_hat)
        grad_input = grad_input.reshape(batchSize,dim,h,w)
```

4.3 Convolutional Neural Networks

```
            return grad_input
class Sqrtm(Function):

    @staticmethod
    def forward(ctx, input, iterN):
        x = input
        batchSize = x.data.shape[0]
        dim = x.data.shape[1]
        dtype = x.dtype
        I3 = 3.0*torch.eye(dim,dim,device = x.device).view(1,dim, dim).repeat(batchSize,1,1).type(dtype)
        normA = (1.0/3.0)*x.mul(I3).sum(dim=1).sum(dim=1)
        A = x.div(normA.view(batchSize,1,1).expand_as(x))
        Y = torch.zeros(batchSize, iterN, dim, dim, requires_grad = False, device = x.device)
        Z = torch.eye(dim,dim,device = x.device).view(1,dim,dim).repeat(batchSize,iterN,1,1)
        if iterN < 2:
            ZY = 0.5*(I3 - A)
            Y[:,0,:,:] = A.bmm(ZY)
        else:
            ZY = 0.5*(I3 - A)
            Y[:,0,:,:] = A.bmm(ZY)
            Z[:,0,:,:] = ZY
            for i in range(1, iterN-1):
                ZY = 0.5*(I3 - Z[:,i-1,:,:].bmm(Y[:,i-1,:,:]))
                Y[:,i,:,:] = Y[:,i-1,:,:].bmm(ZY)
                Z[:,i,:,:] = ZY.bmm(Z[:,i-1,:,:])
            ZY = 0.5*Y[:,iterN-2,:,:].bmm(I3 - Z[:,iterN-2,:,:].bmm(Y[:,iterN-2,:,:]))
        y = ZY*torch.sqrt(normA).view(batchSize, 1, 1).expand_as(x)
        ctx.save_for_backward(input, A, ZY, normA, Y, Z)
        ctx.iterN = iterN
        return y
    @staticmethod
    def backward(ctx, grad_output):
        input, A, ZY, normA, Y, Z = ctx.saved_tensors
        iterN = ctx.iterN
        x = input
        batchSize = x.data.shape[0]
        dim = x.data.shape[1]
        dtype = x.dtype
        der_postCom = grad_output*torch.sqrt(normA).view(batchSize, 1,
```

```
1).expand_as(x)
        der_postComAux = (grad_output*ZY).sum(dim=1).sum(dim=1).
div(2*torch.sqrt(normA))
        I3 = 3.0*torch.eye(dim,dim,device = x.device).view(1,
dim, dim).repeat(batchSize,1,1).type(dtype)
        if iterN < 2:
            der_NSiter = 0.5*(der_postCom.bmm(I3 - A) - A.
bmm(der_sacleTrace))
        else:
            dldY = 0.5*(der_postCom.bmm(I3 - Y[:,iterN-2,:,:].bmm
(Z[:,iterN-2,:,:])) -
                        Z[:,iterN-2,:,:].bmm(Y[:,iterN-2,:,:]).
bmm(der_postCom))
            dldZ = -0.5*Y[:,iterN-2,:,:].bmm(der_postCom).
bmm(Y[:,iterN-2,:,:])
            for i in range(iterN-3, -1, -1):
                YZ = I3 - Y[:,i,:,:].bmm(Z[:,i,:,:])
                ZY = Z[:,i,:,:].bmm(Y[:,i,:,:])
                dldY_ = 0.5*(dldY.bmm(YZ) -
                        Z[:,i,:,:].bmm(dldZ).bmm(Z[:,i,:,:]) -
                            ZY.bmm(dldY))
                dldZ_ = 0.5*(YZ.bmm(dldZ) -
                        Y[:,i,:,:].bmm(dldY).bmm(Y[:,i,:,:]) -
                            dldZ.bmm(ZY))
                dldY = dldY_
                dldZ = dldZ_
            der_NSiter = 0.5*(dldY.bmm(I3 - A) - dldZ - A.
bmm(dldY))
        grad_input = der_NSiter.div(normA.view(batchSize,1,1).
expand_as(x))
        grad_aux = der_NSiter.mul(x).sum(dim=1).sum(dim=1)
        for i in range(batchSize):
            grad_input[i,:,:] += (der_postComAux[i] \
                                - grad_aux[i] / (normA[i] *
normA[i])) \
                    *torch.ones(dim,device = x.device).diag()
        return grad_input, None
class Triuvec(Function):

    @staticmethod
    def forward(ctx, input):
        x = input
        batchSize = x.data.shape[0]
        dim = x.data.shape[1]
        dtype = x.dtype
```

4.3 Convolutional Neural Networks

```python
            x = x.reshape(batchSize, dim*dim)
            I = torch.ones(dim,dim).triu().t().reshape(dim*dim)
            index = I.nonzero()
            y = torch.zeros(batchSize,int(dim*(dim+1)/2),device = x.device)
            for i in range(batchSize):
                y[i, :] = x[i, index].t()
            ctx.save_for_backward(input,index)
            return y
        @staticmethod
        def backward(ctx, grad_output):
            input,index = ctx.saved_tensors
            x = input
            batchSize = x.data.shape[0]
            dim = x.data.shape[1]
            dtype = x.dtype
            grad_input = torch.zeros(batchSize,dim,dim,device = x.device,requires_grad=False)
            grad_input = grad_input.reshape(batchSize,dim*dim)
            for i in range(batchSize):
                grad_input[i,index] = grad_output[i,:].reshape(index.size(),1)
            grad_input = grad_input.reshape(batchSize,dim,dim)
            return grad_input
def CovpoolLayer(var):
    return Covpool.apply(var)
def SqrtmLayer(var, iterN):

    return Sqrtm.apply(var, iterN)

def TriuvecLayer(var):

    return Triuvec.apply(var)

#use

if GSoP_mode == 1:
    self.avgpool = nn.AvgPool2d(14, stride=1)
    self.fc = nn.Linear(512 * block.expansion, num_classes)
    print("GSoP-Net1 generating...")
else :
    self.isqrt_dim = 256
    self.layer_reduce = nn.Conv2d(512 * block.expansion, self.isqrt_dim, kernel_size=1, stride=1, padding=0, bias=False)
    self.layer_reduce_bn = nn.BatchNorm2d(self.isqrt_dim)
```

```
    self.layer_reduce_relu = nn.ReLU(inplace=True)
    self.fc = nn.Linear(int(self.isqrt_dim * (self.isqrt_dim + 1)
/ 2), num_classes)
    print("GSoP-Net2 generating...")
if self.GSoP_mode == 1:

    x = self.avgpool(x)
else :
    x = self.layer_reduce(x)
    x = self.layer_reduce_bn(x)
    x = self.layer_reduce_relu(x)
    x = MPNCOV.CovpoolLayer(x)

    x = MPNCOV.SqrtmLayer(x, 3)
    x = MPNCOV.TriuvecLayer(x)
```

AA-Net

The overall structure of AA-Net is shown in Fig. 4.38. AA-Net employs an attention mechanism that can jointly participate in the spatial and feature subspaces (where each head corresponds to a feature subspace), introducing additional feature maps instead of refining them. The core idea is to utilize the self-attention mechanism. Firstly, the attention weight map is obtained through matrix operations. Then, through multi-head operations, weights are assigned to multiple spaces, and the dot product of attention is performed within these multiple spaces to implement the self-attention mechanism.

AA-Net code implementation based on Pytorch platform is as follows:

```
class AugmentedConv(nn.Module):
    def __init__(self, in_channels, out_channels, kernel_size,
dk, dv, Nh, relative):
        super(AugmentedConv, self).__init__()
        self.in_channels = in_channels
```

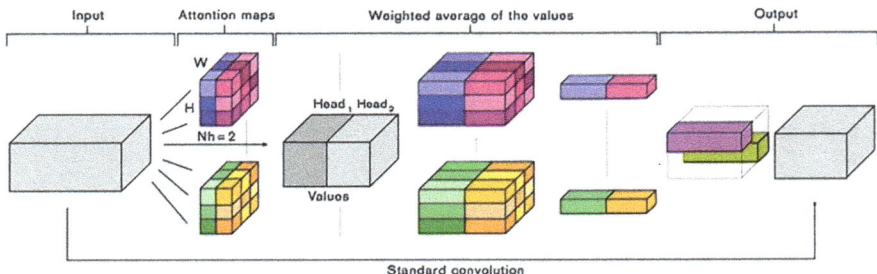

Fig. 4.38 Overall structure of AA-Net

4.3 Convolutional Neural Networks

```python
        self.out_channels = out_channels
        self.kernel_size = kernel_size
        self.dk = dk
        self.dv = dv
        self.Nh = Nh
        self.relative = relative
        self.conv_out = nn.Conv2d(self.in_channels, self.out_channels - self.dv, self.kernel_size, padding=1)
        self.qkv_conv = nn.Conv2d(self.in_channels, 2 * self.dk + self.dv, kernel_size=1)
        self.attn_out = nn.Conv2d(self.dv, self.dv, 1)
    def forward(self, x):

        # Input x
        # (batch_size, channels, height, width)
        batch, _, height, width = x.size(
        # conv_out
        # (batch_size, out_channels, height, width)
        conv_out = self.conv_out(x)
        # flat_q, flat_k, flat_v
        # (batch_size, Nh, height * width, dvh or dkh)
        # dvh = dv / Nh, dkh = dk / Nh
        # q, k, v
        # (batch_size, Nh, height, width, dv or dk)
        flat_q, flat_k, flat_v, q, k, v = self.compute_flat_qkv(x, self.dk, self.dv, self.Nh)
        logits = torch.matmul(flat_q.transpose(2, 3), flat_k)
        if self.relative:
            h_rel_logits, w_rel_logits = self.relative_logits(q)
            logits += h_rel_logits
            logits += w_rel_logits
        weights = F.softmax(logits, dim=-1)
        # attn_out
        # (batch, Nh, height * width, dvh)
        attn_out = torch.matmul(weights, flat_v.transpose(2, 3))
        attn_out = torch.reshape(attn_out, (batch, self.Nh, self.dv / self.Nh, height, width))
        # combine_heads_2d
        # (batch, out_channels, height, width)
        attn_out = self.combine_heads_2d(attn_out)
        attn_out = self.attn_out(attn_out)
        return torch.cat((conv_out, attn_out), dim=1)
```

4.4 Recurrent Neural Network

In baseball, when the batter hits the ball, the catcher immediately starts running and anticipates the trajectory of the ball. The catcher tracked the ball and adjusted its motion, finally catching it to applause. Whether it's a catcher predicting the trajectory of a ball, predicting the winner or loser of a game before a game, or even predicting the taste of coffee in your breakfast, predicting is always what we do. A Recurrent Neural Network (RNN) is a type of neural network that can predict the future. They can analyze time series data (such as stock prices) and then tell you when to buy or sell a stock. In autonomous driving systems, RNN can predict a car's trajectory to help avoid accidents. In general, RNN can work in sequences of any length. For example, RNN can use sentences, documents, or speech samples as input for automatic translation, language conversion to text, or sentiment analysis (such as reading movie reviews, extracting the judges' feelings about the movie) (He 2019).

In addition, the predictive power of RNN also allows them to generate amazing creativity, for example, RNN can be asked to predict what the next note is most likely to appear in the melody, and then randomly select one of the notes and play it. This action is then repeated, continuing to ask the network to predict the next note and play it.

This section mainly involves the following contents: the history and basic concepts of RNN; RNN structure; RNN application instance; recent developments in RNN.

4.4.1 Structure of Recurrent Neural Network

The activation flow of feedforward neural networks is in one direction, from the input layer to the output layer. RNN is very similar to feedforward neural networks except that it has reverse connections. Consider a simple RNN, as shown in Fig. 4.39a, consisting of a single neuron that receives its own input, produces an output, and then returns the output to itself. At each time iterated t, this circulating neuron h receives the input x_t and the previous time iterated its own output o_{t-1}. As

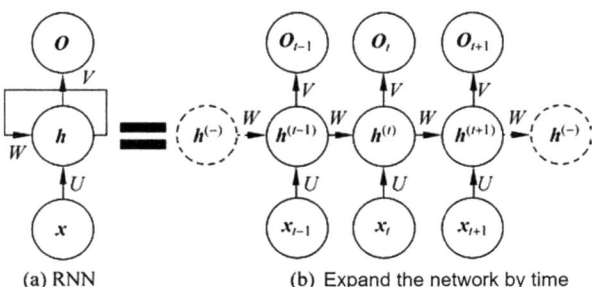

(a) RNN (b) Expand the network by time

Fig. 4.39 Simple RNN

4.4 Recurrent Neural Network

shown in Fig. 4.39b, you can use a timeline to represent the individual tiny networks, which is called expanding the network by time.

Each circulating neuron has two weights: one is the input x_t and the other is the output o_{t-1} of the previous time iteration. Call these two weight vectors w_x and w_o. A single recurrent neural network can be represented as

$$o_t = \phi\left(x_t^T \cdot w_x + o_{t-1}^T \cdot w_o + b\right) \quad (4.25)$$

4.4.2 Application of Recurrent Neural Network and MATLAB Examples

Example 4.7 Suppose that a family uses air conditioning power consumption data for 7 days, and 4 time points are recorded every day. The air conditioning power consumption of the first 3 time points is used to predict the air conditioning power consumption of the fourth time point, and so on. The power consumption of the seventh day is used as the test model.

Solution A simple LSTM network is used to train the data and predict the value of air conditioning power on day 7. MATLAB program implementation code is as follows:

```
%%% LSTM network combined with example simulation
%% Program description
% 1. Data for 7 days, 4 time points of the air conditioning power
consumption, with the first 3 to speculate the fourth training,
and so on, the seventh day as the test
% 2. The LSTM network has 12 input nodes, 4 output nodes, and 18
hidden nodes
clear all;

clc;
%% Data loading and normalization processing
[train_data,test_data]=LSTM_data_process();
data_length=size(train_data,1);
data_num=size(train_data,2);
%% Initialize network parameters
% Node number setting
input_num=12;
cell_num=18;
output_num=4;
% Gate bias in the network
```

```
bias_input_gate=rand(1,cell_num);
bias_forget_gate=rand(1,cell_num);
bias_output_gate=rand(1,cell_num);
% ab=1.2;
% bias_input_gate=ones(1,cell_num)/ab;
% bias_forget_gate=ones(1,cell_num)/ab;
% bias_output_gate=ones(1,cell_num)/ab;
% Network weight initialization
ab=20;
weight_input_x=rand(input_num,cell_num)/ab;
weight_input_h=rand(output_num,cell_num)/ab;
weight_inputgate_x=rand(input_num,cell_num)/ab;
weight_inputgate_c=rand(cell_num,cell_num)/ab;
weight_forgetgate_x=rand(input_num,cell_num)/ab;
weight_forgetgate_c=rand(cell_num,cell_num)/ab;
weight_outputgate_x=rand(input_num,cell_num)/ab;
weight_outputgate_c=rand(cell_num,cell_num)/ab;
% hidden_output weight

weight_preh_h=rand(cell_num,output_num);

% Network status initialization

cost_gate=1e-10;
h_state=rand(output_num,data_num);
cell_state=rand(cell_num,data_num);
%% Network training and learning
for iter=1:4000
    yita=0.15;            % Each iteration weight adjusts the proportion
    for m=1:data_num
        % Feedforward part
        if(m==1)
            gate=tanh(train_data(:,m)'*weight_input_x);

input_gate_input=train_data(:,m)'*weight_inputgate_x+bias_input_gate;

output_gate_input=train_data(:,m)'*weight_outputgate_x+bias_output_gate;
            for n=1:cell_num

input_gate(1,n)=1/(1+exp(-input_gate_input(1,n)));

output_gate(1,n)=1/(1+exp(-output_gate_input(1,n)));
```

4.4 Recurrent Neural Network

```
            end
            forget_gate=zeros(1,cell_num);
            forget_gate_input=zeros(1,cell_num);
            cell_state(:,m)=(input_gate.*gate)';
        else
gate=tanh(train_data(:,m)'*weight_input_x+h_state(:,m-1)'*weight_input_h);

input_gate_input=train_data(:,m)'*weight_inputgate_x+cell_state(:,m-1)'*weight_inputgate_c+bias_input_gate;

forget_gate_input=train_data(:,m)'*weight_forgetgate_x+cell_state(:,m-1)'*weight_forgetgate_c+bias_forget_gate;

output_gate_input=train_data(:,m)'*weight_outputgate_x+cell_state(:,m-1)'*weight_outputgate_c+bias_output_gate;
            for n=1:cell_num

input_gate(1,n)=1/(1+exp(-input_gate_input(1,n)));

forget_gate(1,n)=1/(1+exp(-forget_gate_input(1,n)));

output_gate(1,n)=1/(1+exp(-output_gate_input(1,n)));
            end

cell_state(:,m)=(input_gate.*gate+cell_state(:,m-1)'.*forget_gate)';
        end
        pre_h_state=tanh(cell_state(:,m)').*output_gate;
        h_state(:,m)=(pre_h_state*weight_preh_h)';
        % Error calculation
        Error=h_state(:,m)-test_data(:,m);
        Error_Cost(1,iter)=sum(Error.^2);
        if(Error_Cost(1,iter)<cost_gate)
            flag=1;
            break;
        else
            [   weight_input_x,...
                weight_input_h,...
                weight_inputgate_x,...
                weight_inputgate_c,...
                weight_forgetgate_x,...
                weight_forgetgate_c,...
                weight_outputgate_x,...
```

```
                    weight_outputgate_c,...
weight_preh_h ]=LSTM_updata_weight(m,yita,Error,...
weight_input_x,...
weight_input_h,...
weight_inputgate_x,...
weight_inputgate_c,...
weight_forgetgate_x,...
weight_forgetgate_c,...
weight_outputgate_x,...
weight_outputgate_c,...
weight_preh_h,...
cell_state,h_state,...
input_gate,forget_gate,...
output_gate,gate,...
train_data,pre_h_state,...
input_gate_input,...
output_gate_input,...
forget_gate_input);
        end

    end
    if(Error_Cost(1,iter)<cost_gate)
        break;
    end
end
%% Plot the Error-Cost curve
% for n=1:1:iter
%     text(n,Error_Cost(1,n),'*');
```

4.4 Recurrent Neural Network

```
%       axis([0,iter,0,1]);
%       title('Error-Cost curve');
% end
for n=1:1:iter
    semilogy(n,Error_Cost(1,n),'*');
    hold on;
    title('Error-Cost curve');
end
%% Test using the seventh day data
% Data loading
test_final=[0.4557 0.4790 0.7019 0.8211 0.4601 0.4811 0.7101
0.8298 0.4612 0.4845 0.7188 0.8312]';
test_final=test_final/sqrt(sum(test_final.^2));
test_output=test_data(:,4);
% Feed forward
m=4;
gate=tanh(test_final'*weight_input_x+h_state(:,m-1)'*weight_
input_h);
input_gate_input=test_final'*weight_inputgate_x+cell_
state(:,m-1)'*weight_inputgate_c+bias_input_gate;
forget_gate_input=test_final'*weight_forgetgate_x+cell_
state(:,m-1)'*weight_forgetgate_c+bias_forget_gate;
output_gate_input=test_final'*weight_outputgate_x+cell_
state(:,m-1)'*weight_outputgate_c+bias_output_gate;
for n=1:cell_num
    input_gate(1,n)=1/(1+exp(-input_gate_input(1,n)));
    forget_gate(1,n)=1/(1+exp(-forget_gate_input(1,n)));
    output_gate(1,n)=1/(1+exp(-output_gate_input(1,n)));
end
cell_state_test=(input_gate.*gate+cell_
state(:,m-1)'.*forget_gate)';
pre_h_state=tanh(cell_state_test').*output_gate;
h_state_test=(pre_h_state*weight_preh_h)'
test_output
```

Table 4.3 shows the training effects of examples 4.5.

Table 4.3 Training results

Point of time	True value	Predicted value
1	0.3669	0.3577
2	0.3902	0.3791
3	0.5639	0.5581
4	0.6546	0.6456

4.4.3 Latest Development of Recurrent Neural Networks

RNN has been proved to be very successful in natural language processing, such as word vector expression, sentence validity checking and part-of-speech tagging. Among RNNs, the most widely used and successful model is the LSTM.

LSTM

One of the key points of RNNs is the ability to connect previous information to the current task, for example using past video segments to infer understanding of the current segment. If RNNs can do this, they become very useful.

Sometimes, it is simply necessary to know prior information to perform the current task. For example, a language model is used to predict the next word based on the previous word. If one tries to predict the last word of "the clouds are in the sky", one does not need any other context—the next word should obviously be sky. In such a scenario, the interval between the relevant information and the predicted word position is very small, and the RNN can learn to use the previous information.

But there are also more complex scenarios. Suppose you try to predict the last words of "I grew up in France…. I speak fluent French". The current information suggests that the next word might be the name of a language, but to figure out what language it is would require the context of France, which was mentioned earlier and is far from the current location. This suggests that the gap between the relevant information and the current predicted location must be becoming quite large.

LSTM can deal with such long-term dependency problem. LSTM is a special type of RNN that can learn long-term dependency information. LSTM was proposed by Hochreiter and Schmidhuber, and recently improved and popularized by AlexGraves. LSTM has achieved great success in dealing with many problems and has been widely used.

LSTM avoids the long-term dependency problem by deliberate design. Remembering long-term information is in practice the default behavior of LSTM, not a capability that needs to be acquired at great cost.

All RNNs have a form of chaining of repeating neural network modules. In standard RNNs, this repeating module has only a very simple structure, such as a tanh layer, as shown in Fig. 4.40.

The LSTM also uses this structure, but the repeated modules have different structures. Unlike a single neural network layer, LSTM has four layers that interact in a very specific way, as shown in Fig. 4.41.

LSTM-based systems can perform tasks such as translating languages, controlling robots, image analysis, document summarization, speech recognition image recognition, handwriting recognition, controlling chatbots, predicting diseases, clicks and stocks, synthesizing music, and so on. For example, in 2015, Google dramatically improved speech recognition in Android phones and other devices with an LSTM program trained on CTC. Apple's iPhone uses LSTM in QuickType

4.4 Recurrent Neural Network

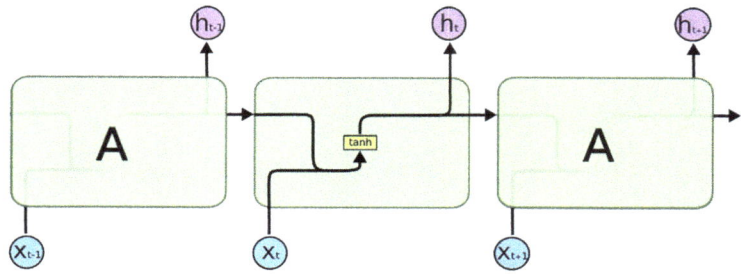

Fig. 4.40 Repetitive modules in a standard RNN contain a single layer

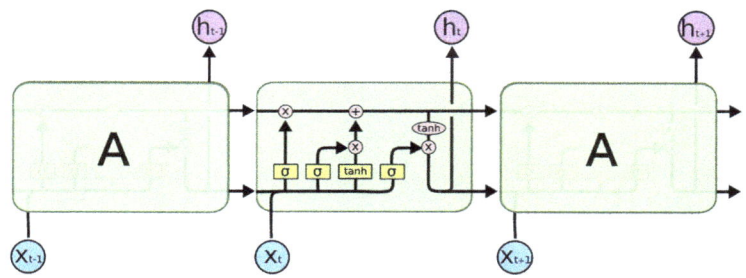

Fig. 4.41 The repeating module in the LSTM contains four interacting layers

and Siri; Microsoft not only uses LSTM for speech recognition, but also uses the technology for virtual dialog image generation and writing program code. Amazon Alexa communicates with you at home through bidirectional LSTM, while Google uses LSTM in a wider range of ways, generating image captions, automatically responding to emails, and more. The use of LSTM in the new smart assistant Allo also significantly improves the quality of Google Translate.

The MATLAB code implementation of the LSTM is as follows:

```
% implementation of LSTM
clc
% clear
close all
%% training dataset generation

binary_dim     = 8;

largest_number = 2^binary_dim - 1;

binary         = cell(largest_number, 1);

for i = 1:largest_number + 1
```

```
    binary{i}      = dec2bin(i-1, binary_dim);
    int2binary{i}  = binary{i};
end
%% input variables

alpha       = 0.1;
input_dim   = 2;
hidden_dim  = 32;
output_dim  = 1;
allErr = [];
%% initialize neural network weights
% in_gate      = sigmoid(X(t) * X_i + H(t-1) * H_i)    ------- (1)
X_i = 2 * rand(input_dim, hidden_dim) - 1;
H_i = 2 * rand(hidden_dim, hidden_dim) - 1;
X_i_update = zeros(size(X_i));
H_i_update = zeros(size(H_i));
bi = 2*rand(1,1) - 1;
bi_update = 0;
% forget_gate = sigmoid(X(t) * X_f + H(t-1) * H_f)    ------- (2)

X_f = 2 * rand(input_dim, hidden_dim) - 1;
H_f = 2 * rand(hidden_dim, hidden_dim) - 1;
X_f_update = zeros(size(X_f));
H_f_update = zeros(size(H_f));
bf = 2*rand(1,1) - 1;
bf_update = 0;
% out_gate     = sigmoid(X(t) * X_o + H(t-1) * H_o)    ------- (3)
X_o = 2 * rand(input_dim, hidden_dim) - 1;
H_o = 2 * rand(hidden_dim, hidden_dim) - 1;
X_o_update = zeros(size(X_o));
H_o_update = zeros(size(H_o));
bo = 2*rand(1,1) - 1;
bo_update = 0;
% g_gate       = tanh(X(t) * X_g + H(t-1) * H_g)       ------- (4)
X_g = 2 * rand(input_dim, hidden_dim) - 1;
H_g = 2 * rand(hidden_dim, hidden_dim) - 1;
X_g_update = zeros(size(X_g));
H_g_update = zeros(size(H_g));
bg = 2*rand(1,1) - 1;
bg_update = 0;
out_para = 2 * rand(hidden_dim, output_dim) - 1;

out_para_update = zeros(size(out_para));
% C(t) = C(t-1) .* forget_gate + g_gate .* in_gate    ------- (5)
% S(t) = tanh(C(t)) .* out_gate                       ------- (6)
```

4.4 Recurrent Neural Network

```
% Out    = sigmoid(S(t) * out_para)                    ------- (7)
% Note: Equations (1)-(6) are cores of LSTM in forward, and equation (7) is
% used to transfer hiddent layer to predicted output, i.e., the output layer.
% (Sometimes you can use softmax for equation (7))
%% train

iter = 99999; % training iterations
for j = 1:iter
    % generate a simple addition problem (a + b = c)
    a_int = randi(round(largest_number/2));   % int version
    a     = int2binary{a_int+1};              % binary encoding
    b_int = randi(floor(largest_number/2));   % int version

    b     = int2binary{b_int+1};              % binary encoding

    % true answer

    c_int = a_int + b_int;                    % int version
    c     = int2binary{c_int+1};              % binary encoding
    % where we'll store our best guess (binary encoded)

    d     = zeros(size(c));
    if length(d)<8
        pause;
    end
    % total error

    overallError = 0;

    % difference in output layer, i.e., (target - out)

    output_deltas = [];

    % values of hidden layer, i.e., S(t)

    hidden_layer_values = [];
    cell_gate_values    = [];
    % initialize S(0) as a zero-vector
    hidden_layer_values = [hidden_layer_values; zeros(1, hidden_dim)];
    cell_gate_values    = [cell_gate_values; zeros(1, hidden_dim)];
    % initialize memory gate
```

```
    % hidden layer
    H = [];
    H = [H; zeros(1, hidden_dim)];
    % cell gate
    C = [];
    C = [C; zeros(1, hidden_dim)];
    % in gate
    I = [];
    % forget gate
    F = [];
    % out gate
    O = [];
    % g gate
    G = [];
    % start to process a sequence, i.e., a forward pass

    % Note: the output of a LSTM cell is the hidden_layer, and you need to
    % transfer it to predicted output
    for position = 0:binary_dim-1
        % X ------> input, size: 1 x input_dim

X = [a(binary_dim - position)-'0' b(binary_dim - position)-'0'];
        % y ------> label, size: 1 x output_dim

        y = [c(binary_dim - position)-'0']';

        % use equations (1)-(7) in a forward pass. here we do not use bias

        in_gate     = sigmoid(X * X_i + H(end, :) * H_i + bi);    % equation (1)
        forget_gate = sigmoid(X * X_f + H(end, :) * H_f + bf);    % equation (2)
        out_gate    = sigmoid(X * X_o + H(end, :) * H_o + bo);    % equation (3)
        g_gate      = tan_h(X * X_g + H(end, :) * H_g + bg);      % equation (4)
        C_t         = C(end, :) .* forget_gate + g_gate .* in_gate;   % equation (5)
        H_t         = tan_h(C_t) .* out_gate;    % equation (6)
        % store these memory gates

        I = [I; in_gate];
```

4.4 Recurrent Neural Network

```
        F = [F; forget_gate];
        O = [O; out_gate];
        G = [G; g_gate];
        C = [C; C_t];
        H = [H; H_t];
        % compute predict output

        pred_out = sigmoid(H_t * out_para);

        % compute error in output layer

        output_error = y - pred_out;

        % compute difference in output layer using derivative

%
output_diff = output_error * sigmoid_output_to_
derivative(pred_out);

output_deltas = [output_deltas; output_error];%*sigmoid_output_
to_derivative(pred_out)];
%
output_deltas = [output_deltas; output_error*(pred_out)];
        % compute total error
        % note that if the size of pred_out or target is 1 x n
or m x n,
        % you should use other approach to compute error. here
the dimension
        % of pred_out is 1 x 1
        overallError = overallError + abs(output_error(1));
        % decode estimate so we can print it out

        d(binary_dim - position) = round(pred_out);
    end
    % from the last LSTM cell, you need a initial hidden layer
difference

    future_H_diff = zeros(1, hidden_dim);

    % stare back-propagation, i.e., a backward pass

    % the goal is to compute differences and use them to
update weights
    % start from the last LSTM cell
```

```
        for position = 0:binary_dim-1
            X = [a(position+1)-'0' b(position+1)-'0'];
            % hidden layer

            H_t = H(end-position, :);          % H(t)
            % previous hidden layer
            H_t_1 = H(end-position-1, :);      % H(t-1)
            C_t = C(end-position, :);          % C(t)
            C_t_1 = C(end-position-1, :);      % C(t-1)
            O_t = O(end-position, :);
            F_t = F(end-position, :);
            G_t = G(end-position, :);
            I_t = I(end-position, :);
            % output layer difference

            output_diff = output_deltas(end-position, :);

            % hidden layer difference

            % note that here we consider one hidden layer is input to both
            % output layer and next LSTM cell. Thus its difference also comes
            % from two sources. In some other method, only one source is taken
            % into consideration.
            % use the equation: delta(l) = (delta(l+1) * W(l+1)) .* f'(z) to
            % compute difference in previous layers. look for more about the
            % proof at http://neuralnetworksanddeeplearning.com/chap2.html
%            H_t_diff = (future_H_diff * (H_i' + H_o' + H_f' + H_g') + output_diff * out_para') ...
%                        .* sigmoid_output_to_derivative(H_t);

H_t_diff = output_diff * (out_para');% .* sigmoid_output_to_derivative(H_t);

%
H_t_diff = output_diff * (out_para') .* sigmoid_output_to_derivative(H_t);
%            future_H_diff = H_t_diff;
%
out_para_diff = output_diff * (H_t) * sigmoid_output_to_
```

4.4 Recurrent Neural Network

```
derivative(out_para);
        out_para_diff = (H_t') * output_diff;         % output layer weight
        % out_gate diference


O_t_diff = H_t_diff .* tan_h(C_t) .* sigmoid_output_to_derivative(O_t);

        % C_t difference


C_t_diff = H_t_diff .* O_t .* tan_h_output_to_derivative(C_t);
%       % C(t-1) difference
%         C_t_1_diff = C_t_diff .* F_t;
        % forget_gate_diffeence


F_t_diff = C_t_diff .* C_t_1 .* sigmoid_output_to_derivative(F_t);
        % in_gate difference


I_t_diff = C_t_diff .* G_t .* sigmoid_output_to_derivative(I_t);
        % g_gate difference


G_t_diff = C_t_diff .* I_t .* tan_h_output_to_derivative(G_t);
        % differences of X_i and H_i

X_i_diff =  X' * I_t_diff;% .* sigmoid_output_to_derivative(X_i);
H_i_diff =  (H_t_1)' * I_t_diff;% .* sigmoid_output_to_derivative(H_i);
        % differences of X_o and H_o

X_o_diff = X' * O_t_diff;% .* sigmoid_output_to_derivative(X_o);
```

```
H_o_diff = (H_t_1)' * O_t_diff;% .* sigmoid_output_to_
derivative(H_o);
        % differences of X_o and H_o


X_f_diff = X' * F_t_diff;% .* sigmoid_output_to_derivative(X_f);

H_f_diff = (H_t_1)' * F_t_diff;% .* sigmoid_output_to_
derivative(H_f);
        % differences of X_o and H_o


X_g_diff = X' * G_t_diff;% .* tan_h_output_to_derivative(X_g);

H_g_diff = (H_t_1)' * G_t_diff;% .* tan_h_output_to_
derivative(H_g);
        % update

        X_i_update = X_i_update + X_i_diff;
        H_i_update = H_i_update + H_i_diff;
        X_o_update = X_o_update + X_o_diff;
        H_o_update = H_o_update + H_o_diff;
        X_f_update = X_f_update + X_f_diff;
        H_f_update = H_f_update + H_f_diff;
        X_g_update = X_g_update + X_g_diff;
        H_g_update = H_g_update + H_g_diff;
        bi_update = bi_update + I_t_diff;
        bo_update = bo_update + O_t_diff;
        bf_update = bf_update + F_t_diff;
        bg_update = bg_update + G_t_diff;
        out_para_update = out_para_update + out_para_diff;
    end
    X_i = X_i + X_i_update * alpha;

    H_i = H_i + H_i_update * alpha;
    X_o = X_o + X_o_update * alpha;
    H_o = H_o + H_o_update * alpha;
    X_f = X_f + X_f_update * alpha;
    H_f = H_f + H_f_update * alpha;
    X_g = X_g + X_g_update * alpha;
    H_g = H_g + H_g_update * alpha;
    bi = bi + bi_update * alpha;
    bo = bo + bo_update * alpha;
    bf = bf + bf_update * alpha;
    bg = bg + bg_update * alpha;
```

4.4 Recurrent Neural Network

```
    out_para = out_para + out_para_update * alpha;
    X_i_update = X_i_update * 0;

    H_i_update = H_i_update * 0;
    X_o_update = X_o_update * 0;
    H_o_update = H_o_update * 0;
    X_f_update = X_f_update * 0;
    H_f_update = H_f_update * 0;
    X_g_update = X_g_update * 0;
    H_g_update = H_g_update * 0;
    bi_update = 0;
    bf_update = 0;
    bo_update = 0;
    bg_update = 0;
    out_para_update = out_para_update * 0;
    if(mod(j,1000) == 0)

        if 1%overallError > 1
            err = sprintf('Error:%s\n', num2str(overallError));
fprintf(err);
        end
        allErr = [allErr overallError];
%        try
            d = bin2dec(num2str(d));
%        catch
%            disp(d);
%        end
        if 1%overallError>1
        pred = sprintf('Pred:%s\n',dec2bin(d,8)); fprintf(pred);
        Tru = sprintf('True:%s\n', num2str(c)); fprintf(Tru);
        end
        out = 0;
        tmp = dec2bin(d,8);
        for i = 1:8
            out = out + str2double(tmp(8-i+1)) * power(2,i-1);
        end
        if 1%overallError>1
        fprintf('%d + %d = %d\n',a_int,b_int,out);
        sep = sprintf('-------%d------\n', j); fprintf(sep);
        end
    end
end
figure;plot(allErr);
function output = sigmoid(x)
    output = 1./(1+exp(-x));
```

```
end
function y = sigmoid_output_to_derivative(output)

    y = output.*(1-output);
end
function y = tan_h_output_to_derivative(x)

    y = (1-x.^2);
end
function y=tan_h(x)
y=(exp(x)-exp(-x))./(exp(x)+exp(-x));
End
```

Word2vec

The concept of NLP itself is so large that it can be divided into "natural language" and "processing". First of all, natural language, different from computer language, is a form of information communication formed in the process of human development, reflecting human thinking, including spoken and written language are expressed in the form of natural language. Now all the languages in the world are natural languages, including Chinese, English, French and so on.

Then we come to "processing". If it is only manual processing, then there is already a special linguistics to carry out related research, and there is no need to emphasize "natural", therefore, this "processing" must be computer processing. But after all, computers are not human beings, can not be like human beings to deal with the text, but need to have their own way of processing. Therefore, natural language processing, simply put, is that the computer receives the user's input in the form of natural language, and internally through the human-defined algorithms for processing, computation and a series of operations, simulating the human understanding of natural language, and return to the user's desired results. Just as machinery liberates human hands, the purpose of natural language processing is to use computers to process large-scale natural language information instead of human beings. It is an intersection of artificial intelligence, computer science and information engineering, involving knowledge of statistics and linguistics. Since language is the proof of human thinking, natural language processing is the highest level of artificial intelligence, and is known as "the pearl in the crown of artificial intelligence" (Russell and Norvig 2016).

In most of the tasks of natural language processing, it is necessary to transfer a large amount of text data into the computer for information mining for subsequent work. However, at present, computers can only deal with numerical values and cannot analyze text directly. Therefore, converting the original text data into numerical data has become a key part of the natural language processing task.

Word2vec is a correlation model used to generate word vectors. These models are shallow two-layer neural networks that are trained to reconstruct linguistic word texts.

4.4 Recurrent Neural Network

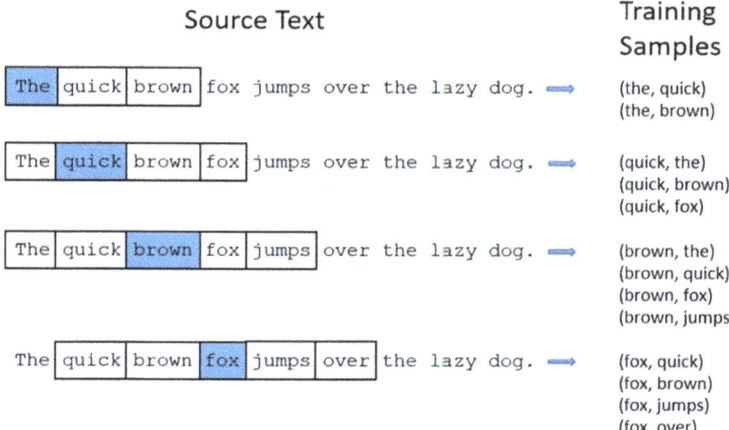

Fig. 4.42 Word2vec training process

In simple terms, Word2vec's series model can convert text (specifically Chinese characters here) into vectors, such as the sentence "I love China", after model processing, may become the following four vectors: (0.12, 0.45, −0.3, 0.44), (0.2, 0.6, 0.7, 0.9), (−0.76, 0.53, 0.88, −0.31), (0.47, 0.92, 0.66, 0.89). This vector is called the word vector (also known as the word vector for Chinese), and the subsequent processing of "I love China" can be converted to the processing of the above four word vectors.

The training idea of the model is generally as follows: At the beginning, a word vector is randomly assigned to each word, and then a word is selected as the central word with a fixed length to obtain the training sample from the original corpus, as shown in Fig. 4.42.

The Python code for Word2vec is as follows:

```
import numpy as np
from collections import defaultdict
class word2vec():

    def __init__(self):

        self.n = settings['n']
        self.lr = settings['learning_rate']
        self.epochs = settings['epochs']
        self.window = settings['window_size']
    def generate_training_data(self, settings, corpus):

        """
        Get training data
        """
```

```python
        #defaultdict(int)  A dictionary that instantiates a
default value with type int when the accessed key does not exist

        word_counts = defaultdict(int)

        #Traverse the corpus

        for row in corpus:
            for word in row:
                #Count the number of times each word appears
                word_counts[word] += 1
        # The length of the vocabulary

        self.v_count = len(word_counts.keys())
        # A list of words in a vocabulary
        self.words_list = list(word_counts.keys())
        # Dictionary data with vocabulary words as key and index
as value
        self.word_index = dict((word, i) for i, word in
enumerate(self.words_list))
        # Dictionary data with index as key and words in the
vocabulary as value
        self.index_word = dict((i, word) for i, word in
enumerate(self.words_list))
        training_data = []

        for sentence in corpus:

            sent_len = len(sentence)

            for i, word in enumerate(sentence):

                w_target = self.word2onehot(sentence[i])

                w_context = []

                for j in range(i - self.window, i + self.window):

                    if j != i and j <= sent_len - 1 and j >= 0:
                        w_context.append(self.
word2onehot(sentence[j]))
                training_data.append([w_target, w_context])

        return np.array(training_data)
```

4.4 Recurrent Neural Network

```python
    def word2onehot(self, word):

        #Encode words with onehot

        word_vec = [0 for i in range(0, self.v_count)]

        word_index = self.word_index[word]

        word_vec[word_index] = 1

        return word_vec

    def train(self, training_data):

        #Randomize parameters w1,w2

        self.w1 = np.random.uniform(-1, 1, (self.v_count, self.n))

self.w2 = np.random.uniform(-1, 1, (self.n, self.v_count))

        for i in range(self.epochs):

            self.loss = 0

            # w_t is a one-hot vector representing the target word

            #w_t -> w_target,w_c ->w_context
            for w_t, w_c in training_data:
                #Forward propagation

                y_pred, h, u = self.forward(w_t)

                #Calculation error

                EI = np.sum([np.subtract(y_pred, word) for word in w_c], axis=0)

                #Backpropagation, update parameters

                self.backprop(EI, h, w_t)

                #Calculate total loss
```

```
            self.loss += -np.sum([u[word.index(1)] for word
in w_c]) + len(w_c) * np.log(np.sum(np.exp(u)))

            print('Epoch:', i, "Loss:", self.loss)

    def forward(self, x):

        """
        Forward propagation
        """
        h = np.dot(self.w1.T, x)

        u = np.dot(self.w2.T, h)

        y_c = self.softmax(u)

        return y_c, h, u

    def softmax(self, x):

        """
        """
        e_x = np.exp(x - np.max(x))
        return e_x / np.sum(e_x)

    def backprop(self, e, h, x):

        dl_dw2 = np.outer(h, e)

        dl_dw1 = np.outer(x, np.dot(self.w2, e.T))

        self.w1 = self.w1 - (self.lr * dl_dw1)

        self.w2 = self.w2 - (self.lr * dl_dw2)

    def word_vec(self, word):

        """

        Get word vector
        Find directly in the weight vector by getting the index
of the word
        """
        w_index = self.word_index[word]
```

4.4 Recurrent Neural Network

```python
            v_w = self.w1[w_index]

            return v_w

    def vec_sim(self, word, top_n):

        """
        Look for similar words
        """
        v_w1 = self.word_vec(word)

        word_sim = {}

        for i in range(self.v_count):

            v_w2 = self.w1[i]
            theta_sum = np.dot(v_w1, v_w2)
            #np.linalg.norm(v_w1) Find the norm, the default is 2 norm, which is the second root of the sum of squares

            theta_den = np.linalg.norm(v_w1) * np.linalg.norm(v_w2)
            theta = theta_sum / theta_den
            word = self.index_word[i]

            word_sim[word] = theta

        words_sorted = sorted(word_sim.items(), key=lambda kv: kv[1], reverse=True)

        for word, sim in words_sorted[:top_n]:

            print(word, sim)

    def get_w(self):

        w1 = self.w1
        return  w1
#hyperparameter
settings = {
    'window_size': 2,    #Window size m
    #The dimension of the word embedding is also the size of the hidden layer.
    'n': 10,
    'epochs': 50,          #Represents the number of times the
```

```
entire sample is traversed. In each epoch, we loop through a
sample of the training set.
    'learning_rate':0.01 #Learning rate
}
#Data preparation

text = "natural language processing and machine learning is fun
and exciting"
#Segmentation of our corpus according to spaces between words
corpus = [[word.lower() for word in text.split()]]
print(corpus)
#Initialize a word2vec object

w2v = word2vec()

training_data = w2v.generate_training_data(settings,corpus)

#Train

w2v.train(training_data)

# Get the vector of words

word = "machine"
vec = w2v.word_vec(word)
print(word, vec)
# Look for similar words

w2v.vec_sim("machine", 3)
```

References

Goodfellow, Ian, Bengio, Yoshua, and Courville, Aaron. Deep Learning. Posts & Telecom Press, 2017.
He, Han. Introduction to Natural Language Processing. Posts & Telecom Press, 2019.
Russell, Stuart, and Norvig, Peter. Artificial Intelligence: A Modern Approach. Posts & Telecom Press, 2016.
Shi, Yan, Han, Liqun, and Lian, Xiaoqin. Design Methods and Case Analyses of Neural Networks. Beijing University of Posts and Telecommunications Press, 2009.
Zelinsky, Richard S. Computer Vision: Algorithms and Applications. China Machine Press, 2015.
Zhu, Daqi, and Shi, Hui. Principles and Applications of Artificial Neural Networks. Science Press, 2006.

Chapter 5
Machine Learning

5.1 Naive Bayes Algorithm

This section will study the Naive Bayes (NB) algorithm in detail. The main contents of this section include:

- The basic principles and ideas of the Naive Bayes algorithm;
- The process of the Naive Bayes algorithm;
- The models of the Naive Bayes algorithm;
- The characteristics and application scenarios of the Naive Bayes algorithm.

5.1.1 The Basic Concept of the Naive Bayes Algorithm

The NB algorithm is one of the most widely-used classification algorithms and one of the few classification algorithms based on probability theory. It is a simplified version of the Bayes algorithm, that is, it assumes that the attribute conditions are independent of each other when the target value is given. Although this simplification reduces the classification performance of the Bayes algorithm to some extent, it greatly simplifies the complexity of the Bayes algorithm in practical application scenarios and is easy to implement. In real-life applications, the Naive Bayes algorithm is widely used in spam filtering, spam classification, credit evaluation, and phishing website detection.

Basic Principle

Naive Bayes Classification (NBC) is a method based on Bayes' theorem and assumes that the feature conditions are independent of each other. First, through the given training set, with the independence of feature words as a pre-assumption, learn the joint probability distribution from input to output, and then based on the learned model, input x to find the output Y with the maximum posterior probability (Ratner 2021).

Given a sample data set $D = \{d_1, d_2, ..., d_n\}$, the corresponding attribute set of sample data is $X = \{x_1, x_2, ..., x_n\}$, and the class variable is $Y = \{y_1, y_2, ..., y_m\}$, then D can be divided into m classes. If $x_1, x_2, ..., x_n$ are independent of each other, then the prior probability of Y is $P_{prior} = P(Y)$, the posterior probability is $P_{post} = P(Y|X)$. According to the Naive Bayes algorithm, the posterior probability can be obtained from the prior probability $P_{prior} = P(Y)$, evidence $P(X)$ and class-conditional probability $P(X|Y)$:

$$P(Y|X) = \frac{P(Y)P(X|Y)}{P(X)} \tag{5.1}$$

Naive Bayes assumes that each feature is independent of each other. In the case where the class is y, eq. (5.1) can be further expressed as:

$$P(X|Y = y) = \prod_{i=1}^{d} P(x_i|Y = y) \tag{5.2}$$

By combining eqs. (5.1) and (5.2), the posterior probability can be obtained:

$$P_{post} = P(Y|X) = \frac{P(Y)\prod_{i=1}^{d} P(x_i|Y = y)}{P(X)} \tag{5.3}$$

Since the value of $P(X)$ is constant, when comparing posterior probabilities, only the molecular part of eq. (5.3) needs to be compared. Thus, the Naive Bayes of a sample data belonging to class y_i can be obtained:

$$P(y_i|x_1, x_2, ..., x_d) = \frac{P(y_i)\prod_{i=1}^{d} P(x_i|y_i)}{\prod_{i=1}^{d} P(y_i)} \tag{5.4}$$

Basic Ideas

The basic principles of the Naive Bayes algorithm were introduced above. Below, the basic ideas of the Naive Bayes algorithm will be understood through a simple example.

Example 5.1 A hospital received 6 patients in the morning, and now the seventh patient is a construction worker with a sneeze. What is the probability that he has a cold (assuming that the two features of "symptom" and "occupation" are independent of each other)? The patient's condition is shown in Table 5.1.

5.1 Naive Bayes Algorithm

Table 5.1 Patient condition comparison table

Symptom	Occupation	Disease
Sneeze	Nurse	Cold
Sneeze	Farmer	Allergy
Headache	Construction Worker	Concussion
Headache	Construction Worker	Cold
Sneeze	Teacher	Cold
Headache	Teacher	Concussion

Solution According to Bayes' theorem

$$P(Y|X) = \frac{P(Y)P(X|Y)}{P(x)}$$

It can be obtained that:

$$P(cold|sneeze \times construction\ worker)$$
$$= \frac{P(sneeze \times construction\ worker|cold) \times P(cold)}{P(sneeze \times construction\ worker)}$$

It is assumed that the two features of "sneezing" and "construction worker" are independent, so:

$$P(cold|sneeze \times construction\ worker)$$
$$= \frac{P(sneezing|cold) \times P(contruction\ worker|cold) \times P(cold)}{P(sneezing) \times P(construction\ worker)}$$

Then

$$P(cold|sneezing \times construction\ worker) = \frac{\frac{2}{3} \times \frac{1}{3} \times \frac{1}{2}}{\frac{1}{3} \times \frac{1}{2}} = \frac{2}{3}$$

Therefore, this construction worker who sneezes has a 66% probability of having a cold. Similarly, the probabilities of this patient having an allergy or a concussion can be calculated. By comparing these probabilities, we can know what disease the patient is most likely to have.

This is the basic method of the Naive Bayes algorithm, that is, based on probability, calculating the probabilities of various classes according to some features, thereby achieving classification.

5.1.2 The Process and Model of the Naive Bayes Algorithm

Specific Process

The Naive Bayes algorithm is divided into three stages, and the specific process is shown in Fig. 5.1.

The first stage—the preparation stage. According to the specific situation, determine the feature attributes, and make appropriate divisions for each feature attribute. Then, through manual methods, some categories are classified to form a training sample set. The input of this stage is all the data to be classified, and the output is the feature attributes and training samples. This stage is the only stage in the entire Naive Bayes classification that requires manual work. Its quality has an important influence on the entire process. The quality of the classifier is largely determined by the feature attributes and the classification methods.

The second stage—the Naive Bayes classification learning stage. The task of this stage is to generate a classifier. The main work is to calculate the occurrence frequency of each category in the training sample and estimate the conditional probability of each feature attribute for each category, and record the results. The input is the feature attributes and training samples, and the output is the classifier. This stage is a mechanical stage, and the formula discussed above can be automatically calculated by the program.

The third stage—the prediction stage. The task of this stage is to use the classifier to classify the categories to be classified. The input is the categories to be classified,

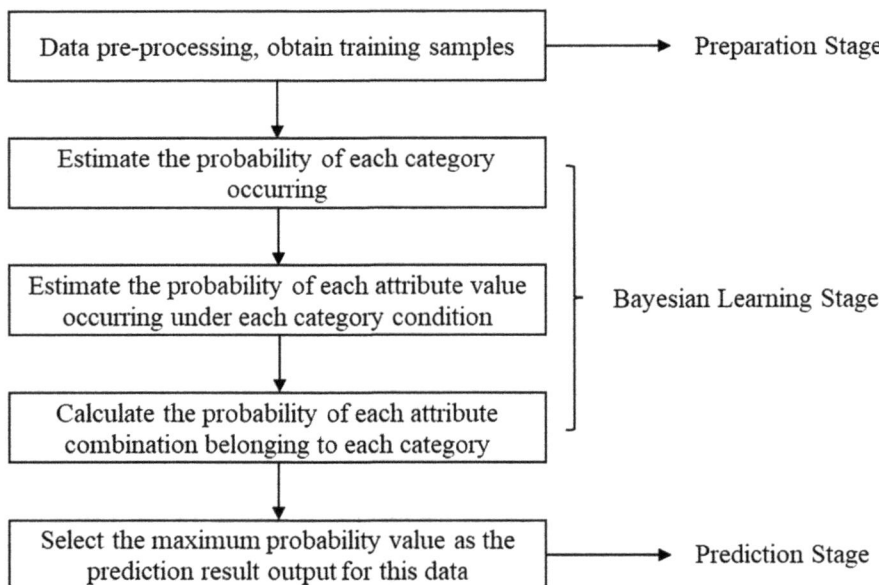

Fig. 5.1 The process flow of the Naive Bayes algorithm

5.1 Naive Bayes Algorithm

and the output is the correspondence between the categories to be classified and the categories. This stage is also a mechanical stage, and is completed by the program.

Commonly Used Models

The Naive Bayes algorithm has three commonly used models (Wei 2018).

1. Gaussian Model: It is used to handle continuous—type variables. When the feature is a continuous—type variable, using the polynomial model will lead to many $P(x_i|y_j) = 0$ (in the case of not being smooth). Even if smoothing is done, the obtained conditional probability is also difficult to describe the real situation. Therefore, for continuous—type variables, the Gaussian model should be used.
2. Polynomial Model: It is the most common and requires that the features are discrete data. When the feature is discrete data, the polynomial model is used. When calculating the prior probability $P(y_k)$ and the conditional probability $P(x_i|y_j)$ some smoothing processing will be done. If smoothing is not done, when a certain feature value xi does not appear in the training sample ($P(x_i|y_j) = 0$), it will lead to the posterior probability being 0. By adding smoothing, this problem can be solved.
3. Bernoulli Model: It requires that the features are discrete, and are Boolean-type, that is, true and false, or 1 and 0. Similar to the polynomial model, the Bernoulli model is suitable for discrete-type features. However, in the Bernoulli model, the value that each feature can take can only be 1 and 0 (in text classification, for example, when a certain word appears in the document, its feature value is 1, otherwise it is 0). In the Bernoulli model, the calculation method of the conditional probability $P(x_i|y_j)$ is

$$\begin{cases} P(x_i|y_k) = P(x_i = 1|y_k), \ x_i = 1 \\ P(x_i|y_k) = 1 - P(x_i = 1|y_k), x_i = 0 \end{cases}$$

5.1.3 Characteristics and Application Scenarios of the Naive Bayes Algorithm

Algorithm Characteristics

The Naive Bayes algorithm assumes that the attributes of the data set are mutually independent, so the logic of the algorithm is very simple and the algorithm is relatively stable. When the data presents different features, the classification performance of the Naive Bayes will not have much difference. In other words, the robustness of the Naive Bayes algorithm is relatively good, and it will not show too much difference for different types of data sets. When the relationship between the attributes of the data set is relatively independent, the Naive Bayes classification algorithm will have a better effect.

The condition of attribute independence is also the deficiency of the Naive Bayes classifier. The independence of data set attributes is difficult to satisfy in many cases, because there are often correlations between the attributes of the data set. If this problem occurs in the classification process, it will lead to a significant reduction in classification efficiency.

Actual Application Scenarios

The Naive Bayes algorithm is widely used, mainly in the following aspects:
1. Text intelligent classification.
2. Spam email filtering.
3. Patient classification.
4. Spelling check.
5. Credit evaluation.
6. Phishing website detection.

5.1.4 Relevant Applications of the Naive Bayes Algorithm and MATLAB Examples

Application Example 1: Judging Whether There Is an Object

Suppose a data set is given, and based on this data set, for a new decision vector, based on formulas (5.1) to (5.4), a maximum probability value can be obtained, and the maximum probability event is output as the target value, which is the classification of this decision vector.

Example 5.2 Taking the data in Table 5.2 as an example, where 1 represents yes and 0 represents no, now knowing that a person has a nice voice, low appearance level, low EQ, and high IQ ($X_1 = 1$, $X_2 = 0$, $X_3 = 0$, $X_4 = 1$), can it be judged whether he has (will have) a partner?

Table 5.2 Comparison table

X_1: Have a nice voice	X_2: Have high appearance value	X_3: Have high emotional intelligence	X_4: Have high intelligence quotient	Y: Can find a partner
1	0	0	0	0
0	1	1	0	1
0	0	1	1	1
1	0	1	0	1
0	1	1	0	1
1	1	1	1	0
0	0	0	1	0

5.1 Naive Bayes Algorithm

Solution $P(X_i|Y=1) = \left[\dfrac{1}{4}, \dfrac{2}{4}, 0, \dfrac{1}{4}\right]$

$$P(X_i|Y=0) = \left[\dfrac{2}{3}, \dfrac{2}{3}, \dfrac{2}{3}, \dfrac{2}{3}\right]$$

$$P(Y=1) = \dfrac{4}{7}$$

$$P(Y=0) = \dfrac{3}{7}$$

The probability of having a partner is $P = 0$, and the probability of not having a partner is $P = 48/576 = 0.0847$, indicating that this person is very likely not to have a partner. The MATLAB code is as follows:

```
clc;clear all;
input = load("BayesData.txt")
[l,w] = size(input);
count = zeros(2,w);
for i = 1:1:l
    for j = 1:1:w
        if input(i,j) == 1 && input(i,end) == 1
            count(1,j) = count(1,j) + 1;
        elseif input(i,j) == 1 && input(i,end) == 0
            count(2,j) = count(2,j) + 1;
        end
    end
end
end
count(2,end) = 1 - count(1,end);
test_data = [1 0 0 1];
answer = [0,0];
% case 1:
temp = 1;
for i = 1:1:w - 1
    if test_data(i) == 1
        temp = temp * count(1,i)/count(1,end);
    else
        temp = temp * (1 - count(1,i)/count(1,end));
    end
end
answer(1) = count(1,end)/l * temp;
% case 0:
temp = 1;
for i = 1:1:w - 1
```

```
        if test_data(i) == 1
            temp = temp * count(2,i)/count(2,end);
        else
            temp = temp * (1 - count(2,i)/count(2,end));
        end
end
answer(2) = count(2,end)/l * temp;
if answer(1) > answer(2)
    disp("May have a partner.")
else
    disp("May not have a partner.")
end
```

The implementation result is shown in Fig. 5.2.

From the result of Example 5.2, it can be seen that a person with a nice voice and high IQ, but with a low—looking appearance and low EQ, still has difficulty in finding a partner.

Application Example 2: Judging the Quality of a Watermelon

Example 5.3 There is a watermelon now, and its attribute values are as follows:

Color: green; peduncle: curly; sound when knocked: dull; texture: clear;

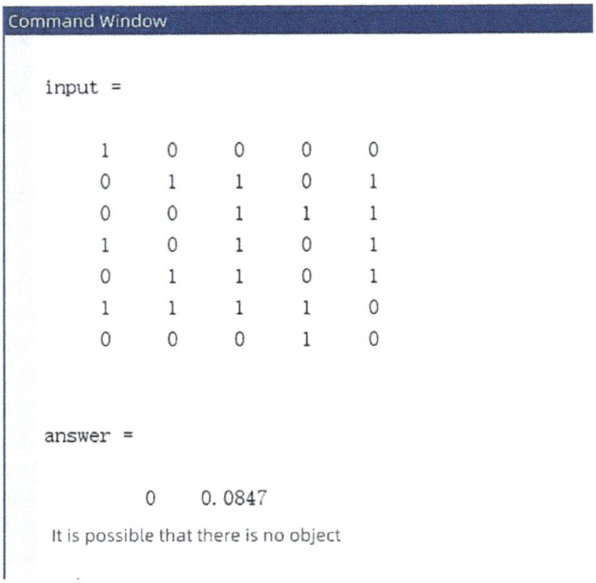

```
Command Window

    input =

        1    0    0    0    0
        0    1    1    0    1
        0    0    1    1    1
        1    0    1    0    1
        0    1    1    0    1
        1    1    1    1    0
        0    0    0    1    0

    answer =

             0    0.0847
    It is possible that there is no object
```

Fig. 5.2 The implementation results of MATLAB code

5.1 Naive Bayes Algorithm

Navel: sunken; touch: hard and slippery; density: 0.697; sugar content: 0.460.

The comparison table of watermelon attributes is shown in Table 5.3. Determine whether this watermelon is good or bad.

Solution First, find the prior probability of each class, that is, the proportion of good watermelons and bad watermelons.

$$P(good\ watermelon) = \frac{8}{17} = 0.471$$

$$P(bad\ watermelon) = \frac{9}{17} = 0.529$$

Then estimate the probability for each attribute value:

$$P(color = green|good\ watermelon) = \frac{3}{8} = 0.375$$

$$P(color = green|bad\ watermelon) = \frac{3}{9} = 0.333$$

$$P(peduncle = curly|good\ watermelon) = \frac{5}{8} = 0.625$$

$$P(peduncle = curly|bad\ watermelon) = \frac{3}{9} = 0.333$$

$$P(sound\ when\ knocked = dull|good\ watermelon) = \frac{6}{8} = 0.750$$

$$P(sound\ when\ knocked = dull|bad\ watermelon) = \frac{4}{9} = 0.444$$

$$P(texture = clear|good\ watermelon) = \frac{7}{8} = 0.875$$

$$P(texture = clear|bad\ watermelon) = \frac{2}{9} = 0.222$$

$$P(navel = sunken|good\ watermelon) = \frac{6}{8} = 0.750$$

$$P(navel = sunken|bad\ watermelon) = \frac{2}{9} = 0.222$$

Table 5.3 Watermelon attribute comparison table

Number	Color	Root	Knocking Sound	Texture	Belly	Touch	Density	Sugar Content	Good Melon
1	Dark green	Curled	Dull sound	Clear	Sunken	Hard and slippery	0.697	0.460	Yes
2	Black	Curled	Dull sound	Clear	Sunken	Hard and slippery	0.774	0.376	Yes
3	Black	Curled	Dull sound	Clear	Sunken	Hard and slippery	0.634	0.364	Yes
4	Dark green	Curled	Dull sound	Clear	Sunken	Hard and slippery	0.608	0.318	Yes
5	Light white	Curled	Dull sound	Clear	Sunken	Hard and slippery	0.556	0.215	Yes
6	Dark green	Slightly curled	Dull sound	Clear	Slightly sunken	Soft and sticky	0.403	0.237	Yes
7	Black	Slightly curled	Dull sound	Slightly blurry	Slightly sunken	Soft and sticky	0.481	0.149	Yes
8	Black	Slightly curled	Dull sound	Clear	Slightly sunken	Hard and slippery	0.437	0.211	Yes
9	Black	Slightly curled	Dull sound	Slightly blurry	Slightly sunken	Hard and slippery	0.666	0.091	No
10	Dark green	Stiff	Crisp	Clear	Flat	Soft and sticky	0.243	0.267	No
11	Light white	Stiff	Crisp	Blurry	Flat	Hard and slippery	0.245	0.057	No
12	Light white	Curled	Dull sound	Blurry	Flat	Soft and sticky	0.343	0.099	No
13	Dark green	Slightly curled	Dull sound	Slightly blurry	Sunken	Hard and slippery	0.639	0.161	No
14	Light white	Slightly curled	Dull sound	Slightly blurry	Sunken	Hard and slippery	0.657	0.198	No
15	Black	Slightly curled	Dull sound	Clear	Slightly sunken	Soft and sticky	0.360	0.370	No
16	Light white	Curled	Dull sound	Blurry	Flat	Hard and slippery	0.593	0.042	No
17	Dark green	Curled	Dull sound	Slightly blurry	Slightly sunken	Hard and slippery	0.719	0.103	No

5.1 Naive Bayes Algorithm

$$P(touch = hard\ and\ slippery | good\ watermelon) = \frac{6}{8} = 0.750$$

$$P(touch = hard\ and\ slippery | bad\ watermelon) = \frac{6}{9} = 0.667$$

$$P(density = 0.697 | good\ watermelon) = \frac{1}{\sqrt{2\pi} \times 0.129} e^{-\frac{(0.697-0.627)^2}{2 \times 0.129^2}} = 1.959$$

$$P(density = 0.697 | bad\ watermelon) = \frac{1}{\sqrt{2\pi} \times 0.195} e^{-\frac{(0.697-0.627)^2}{2 \times 0.129^2}} = 1.203$$

$$P(sugar\ content = 0.460 | good\ watermelon) = \frac{1}{\sqrt{2\pi} \times 0.101} e^{-\frac{(0.460-0.495)^2}{2 \times 0.101^2}} = 0.788$$

$$P(sugar\ content = 0.460 | good\ watermelon) = \frac{1}{\sqrt{2\pi} \times 0.108} e^{-\frac{(0.460-0.495)^2}{2 \times 0.108^2}} = 0.066$$

Calculate the probability that the watermelon is good or bad. The larger probability indicates which type the watermelon is.

$$P(good\ watermelon) \times P(color = green | good\ watermelon)$$
$$P(peduncle = curly | good\ watermelon) = 0.038$$

$$P(bad\ watermelon) \times P(color = green | bad\ watermelon)$$
$$P(peduncle = curly | bad\ watermelon) = 6.80 \times 10^{-5}$$

From this, it can be seen that this watermelon is good.

In the implementation process, the data is divided into training sets and test sets. Calculate the prior probability of each class in the training set (i.e., the proportion of each class in the training set), and then estimate the conditional probability for each attribute of the sample (i.e., the proportion of samples with the same attribute value in each class). The MATLAB code is as follows:

```
[b] = xlsread('mix.xls',1,'A1:C1628');
x = b(:,1);
y = b(:,2);
z = b(:,3);
data = [x,y,z];
NUM = 500; % sample size
Test = sortrows([x(1:NUM,1),y(1:NUM,1 ,z(1:NUM,1)],3); % sort the
samples by class
temp = zeros(2,5); % store the mean, variance and probability of
each class
```

```
% calculate the mean, variance and probability of each class
for i = 1:2
    X = [];
    Y = [];
    count = 0;
    for j = 1:NUM
        if Test(j,3) == i
            X = [X;Test(j,1)];
            Y = [Y;Test(j,2)];
            count = count + 1;
        end
    end
    temp(i,1) = mean(X);
    temp(i,2) = std(X);
    temp(i,3) = mean(Y);
    temp(i,4) = std(Y);
    temp(i,5) = count/NUM;
end
% calculate the prediction results
result = [];
for m = 1:1628
    pre = [];
    for n = 1:2
        PX = 1/(temp(n,2)*sqrt(2*pi))*exp(-((data(m,1)-temp(n,1))^2)/(2*(temp(n,2)^2)));
        PY = 1/(temp(n,4)*sqrt(2*pi))*exp(-((data(m,2)-temp(n,3))^2)/(2*(temp(n,4)^2)));
        pre = [pre;PX*PY*temp(n,5)*10^8];
    end
    [da,index] = max(pre);
    result = [result;index];
end
xlswrite('mix.xls',result,'E1:E1628');
% draw the graph
for i = 1:1628
    rand('seed',result(i,1));
    color = rand(1,3);
    plot(x(i,1),y(i,1),'.','color',color);
    hold on;
end
% Check the accuracy rate
num = 0;
for i = 1:1628
if result(i) == c(i)
num = num + 1; % The number of correct results
end
end
```

5.1 Naive Bayes Algorithm

The implementation result is shown in Fig. 5.3.

Application Example 3: Judging the Category of Flowers

Example 5.4 Suppose there are three types of flowers, and their quantities in nature are the same, that is, randomly picking a flower from these three types, $P(A) = P(B) = P(C) = 1/3$. Now there is a flower, and determine which category it belongs to. Without any prompting conditions, it can be known that the possibilities of belonging to the three types of flowers are the same. If at this time, a 4-dimensional vector x is used: the length of the calyx, the width of the calyx, the length of the petal, and the width of the petal represent their respective features, and these features are known, then determine which category it belongs to.

Solution Given the features of a sample, determining which category it belongs to is the task of pattern recognition. Given the features of a sample, calculating the probabilities that it belongs to these categories, and the largest one is the category it belongs to, which is the method of Bayes classification. Take A as an example, using the Bayes formula:

$$P(A|x) = \frac{P(A,x)}{P(x)} = \frac{P(x|A)P(A)}{P(x)} = \frac{P(x|A)P(A)}{\sum_{i=1}^{3} P(x|\omega_i)P(\omega_i)}, \omega_i = A, B, C$$

Among them, $P(x)$ represents the overall density distribution of the length of the calyx, the width of the calyx, the length of the petal, and the width of the petal for these three types of flowers, which are all the same. $P(A) = 1/3$ is called the prior probability, which is known in practice. $P(x|A)$ is the class-conditional density, that

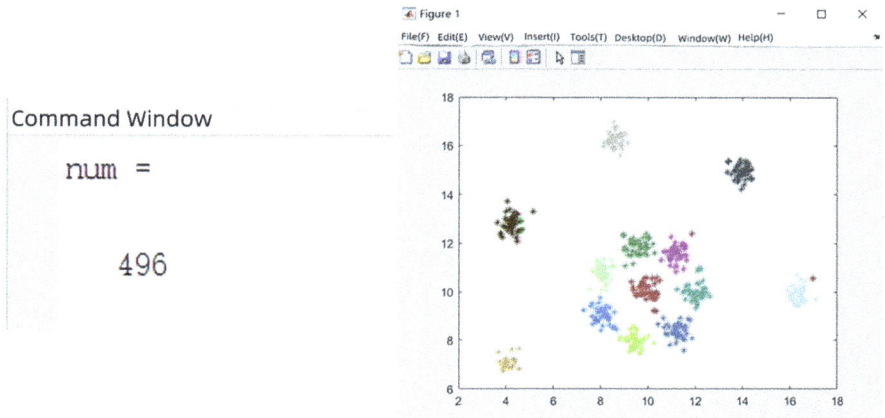

Fig. 5.3 The result graph of Matlab code implementation

is, the distribution of the length of the calyx, the width of the calyx, the length of the petal, and the width of the petal of type flowers. In Naive Bayes classification, assume that this distribution density is a 4-element Gaussian distribution; $P(A|x)$ is called the posterior probability. Therefore, when solving $P(A|x)$, only the numerator of its expanded formula needs to be solved.

The MATLAB code is as follows:

```
clear;clc;
A=[5.1,3.5,1.4,0.2
4.9,3.0,1.4,0.2
4.7,3.2,1.3,0.2
4.6,3.1,1.5,0.2
5.0,3.6,1.4,0.2
5.4,3.9,1.7,0.4
4.6,3.4,1.4,0.3
5.0,3.4,1.5,0.2
4.4,2.9,1.4,0.2
4.9,3.1,1.5,0.1
5.4,3.7,1.5,0.2
4.8,3.4,1.6,0.2
4.8,3.0,1.4,0.1
4.3,3.0,1.1,0.1
5.8,4.0,1.2,0.2
5.7,4.4,1.5,0.4
5.4,3.9,1.3,0.4
5.1,3.5,1.4,0.3
5.7,3.8,1.7,0.3
5.1,3.8,1.5,0.3
5.4,3.4,1.7,0.2
5.2,4.1,1.5,0.1
5.5,4.2,1.4,0.2
4.9,3.1,1.5,0.1
5.0,3.2,1.2,0.2
5.5,3.5,1.3,0.2
4.9,3.1,1.5,0.1
4.4,3.0,1.3,0.2
5.1,3.4,1.5,0.2
5.0,3.5,1.3,0.3
4.5,2.3,1.3,0.3
4.4,3.2,1.3,0.2
5.0,3.5,1.6,0.6
5.1,3.8,1.9,0.4
4.8,3.0,1.4,0.3
5.1,3.8,1.6,0.2
4.6,3.2,1.4,0.2
5.3,3.7,1.5,0.2
5.0,3.3,1.4,0.2
7.0,3.2,4.7,1.4] ;

B=[6.4,3.2,4.5,1.5
6.9,3.1,4.9,1.5
5.5,2.3,4.0,1.3
6.5,2.8,4.6,1.5
5.7,2.8,4.5,1.3
6.3,3.3,4.7,1.6
4.9,2.4,3.3,1.0
6.6,2.9,4.6,1.3
5.2,2.7,3.9,1.4
5.0,2.0,3.5,1.0
5.9,3.0,4.2,1.5
6.0,2.2,4.0,1.0
6.1,2.9,4.7,1.4
5.6,2.9,3.6,1.3
6.7,3.1,4.4,1.4
5.6,3.0,4.5,1.5
5.8,2.7,4.1,1.0
6.2,2.2,4.5,1.5
5.6,2.5,3.9,1.1
5.9,3.2,4.8,1.8
6.1,2.8,4.0,1.3
6.3,2.5,4.9,1.5
6.1,2.8,4.7,1.2
6.4,2.9,4.3,1.3
6.6,3.0,4.4,1.4
6.8,2.8,4.8,1.4
6.7,3.0,5.0,1.7
6.0,2.9,4.5,1.5
5.7,2.6,3.5,1.0
5.5,2.4,3.8,1.1
5.5,2.4,3.7,1.0
5.8,2.7,3.9,1.2
6.0,2.7,5.1,1.6
5.4,3.0,4.5,1.5
6.0,3.4,4.5,1.6
6.7,3.1,4.7,1.5
6.3,2.3,4.4,1.3
5.6,3.0,4.1,1.3
5.7,2.8,4.1,1.3];

C=[6.3,3.3,6.0,2.5
5.8,2.7,5.1,1.9
7.1,3.0,5.9,2.1
6.3,2.9,5.6,1.8
6.5,3.0,5.8,2.2
7.6,3.0,6.6,2.1
4.9,2.5,4.5,1.7
7.3,2.9,6.3,1.8
6.7,2.5,5.8,1.8
7.2,3.6,6.1,2.5
6.5,3.2,5.1,2.0
6.4,2.7,5.3,1.9
6.8,3.0,5.5,2.1
5.7,2.5,5.0,2.0
5.8,2.8,5.1,2.4
6.4,3.2,5.3,2.3
6.5,3.0,5.5,1.8
7.7,3.8,6.7,2.2
7.7,2.6,6.9,2.3
6.0,2.2,5.0,1.5
6.9,3.2,5.7,2.3
5.6,2.8,4.9,2.0
7.7,2.8,6.7,2.0
6.3,3.4,5.6,2.4
6.4,3.1,5.5,1.8
6.0,3.0,4.8,1.8
6.9,3.1,5.4,2.1
6.7,3.1,5.6,2.4
6.9,3.1,5.1,2.3
5.8,2.7,5.1,1.9
6.8,3.2,5.9,2.3
6.7,3.3,5.7,2.5
6.7,3.0,5.2,2.3
6.3,2.5,5.0,1.9
6.5,3.0,5.2,2.0
6.2,3.4,5.4,2.3
5.9,3.0,5.1,1.8];
```

5.1 Naive Bayes Algorithm

```
NA=size(A,1);NB=size(B,1);NC=size(C,1);
A_train=A(1:floor(NA/2),:); % Training data takes 1/2 (or 1/3,
3/4, 1/4)
B_train=B(1:floor(NB/2),:);
C_train=C(1:floor(NC/2),:);
u1=mean(A_train)';u2=mean(B_train)';u3=mean(C_train)';
S1=cov(A_train);S2=cov(B_train);S3=cov(C_train);
S11=inv(S1);S22=inv(S2);S33=inv(S3);
S1_d=det(S1);S2_d=det(S2);S3_d=det(S3);
PA=1/3;PB=1/3;PC=1/3; % Assuming prior probabilities of each
class are equal, i.e., all are 1/3
A_test=A((floor(NA/2)+1):end,:);
B_test=B((floor(NB/2)+1):end,:);
C_test=C((floor(NC/2)+1):end,:);
%test of Sample_A
right1=0;
error1=0;
for i=1:size(A_test,1)
P1=(-1/2)*(A_test(i,:)'-u1)'*S11*(A_test(i,:)'-u1)-
(1/2)*log(S1_d)+log(PA);
P2=(-1/2)*(A_test(i,:)'-u2)'*S22*(A_test(i,:)'-u2)-
(1/2)*log(S2_d)+log(PB);
P3=(-1/2)*(A_test(i,:)'-u3)'*S33*(A_test(i,:)'-u3)-
(1/2)*log(S3_d)+log(PC);
P=[P1 P2 P3];
[Pm,ind]=max(P);
if ind==1
right1=right1+1;
else
error1=error1+1;
end
end
right_rate=right1/size(A_test,1) % Calculate the accuracy rate of
test data in A, and similarly, tests can be done for B, C
```

5.2 Decision Trees

This section will study decision trees (Decision Tree, DT) in detail. The main contents of this section include:

- The definition and characteristics of decision trees;
- The selection of decision tree splitting attributes;
- The conditions for decision trees to stop splitting;

- The pruning of decision trees;
- The three algorithms of decision trees;
- The steps of decision tree generation algorithms.

5.2.1 The Basic Concept of Decision Trees

Decision trees are a commonly-used classification method. The decision tree model is simple and easy to understand, and decision trees can be converted into SQL statements very easily, which can effectively access databases. In particular, in many cases, the accuracy of decision tree classifiers is similar to that of other classification methods, or even better. Currently, several decision tree algorithms have been developed, such as ID3, CART, C4.5, SLIQ, SPRINT, etc. This section mainly introduces the definition and characteristics of decision trees (Wang 2018).

Definition of Decision Trees

A decision tree is based on the probability of occurrence of various situations. By constructing a decision tree to obtain the expected value of net present value, the expected value is greater than or equal to the zero-profit point, evaluating project risks, and analyzing the decision-making rationality of their feasibility. It is a graphical method for directly using probability analysis. Because the decision-making branch chart drawn is very much like a tree-shaped trunk, it is called a decision tree.

A decision tree is a tree-shaped structure. Each internal node represents a test on an attribute, each branch represents an output of the test, and each leaf node represents a class. Decision trees are a very common classification method.

Example 5.5 The manager of a golf club aims to adjust the number of club employees to reduce capital waste by predicting when people will play golf through next-week weather forecasts, so as to adjust the number of employees in a timely manner.

Solution By collecting weather conditions (sunny, cloudy, and rainy), relative humidity (percentage), presence of wind, and whether customers are at the club on these days, a data table with 14 columns and 5 rows is finally obtained. Based on the above independent variables, the decision tree shown in Fig. 5.4 is established.

Similar to the human thinking process, the decision-making process is like a tree growing from a root (the tree is growing vertically, and it may also grow horizontally). The top-most node is called the root node, and the node where the judgment result is placed at the bottom is called the leaf node or the end node. The descriptive decision tree is multi-branched, that is, each non-leaf node has two or three branches. The variable on the decision tree node may be in various forms (continuous, discrete, ordered, classified variables, etc.), and a variable can also repeat at different

5.2 Decision Trees

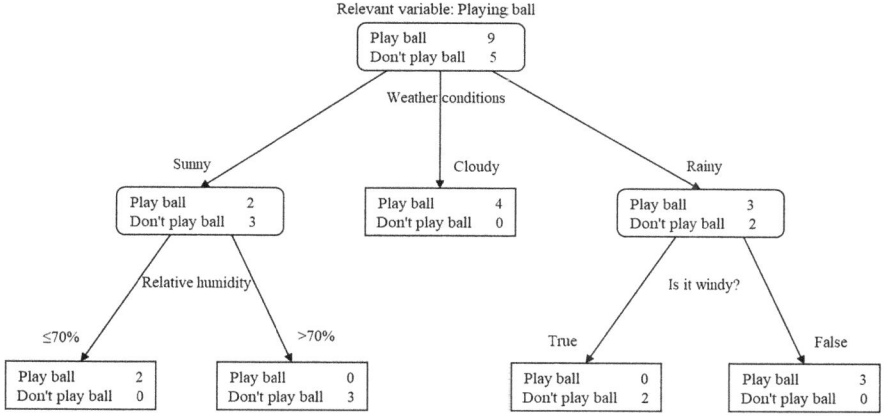

Fig. 5.4 Example of a decision tree

nodes. The node in front of a node is called the parent node (parent node or parent node), and the node corresponding to the child node of the front node is called the child node (child node or child node), and the nodes listed side by side are called sibling nodes (sibling nodes).

Characteristics of Decision Trees

The advantages of decision trees are as follows:

1. Decision trees are easy to understand and implement, and can directly reflect the characteristics of data. People do not need to know a lot of background knowledge during the learning process, and can understand the meaning expressed by the decision tree through necessary explanations.
2. For decision trees, data preparation is usually simple or unnecessary, and can simultaneously process data types and regular-type attributes, and can obtain feasible and effective results for large-scale data sources in a relatively short time.
3. It is easy to evaluate the model through static testing, and the reliability of the model can be measured. If a given observation model is given, the corresponding logical expression can be easily derived from the generated decision tree.

The disadvantages of decision trees are as follows:

1. It is relatively difficult to predict continuous fields.
2. For data with time sequence, a lot of pre-processing work is required.
3. When there are too many categories, errors may increase faster.
4. When classifying algorithms, generally only one field is used for classification.

5.2.2 Construction of Decision Trees

The construction of decision trees is a process of step-by-step data splitting. The construction steps are as follows:

Step 1: Treat all data as a node and enter Step 2.

Step 2: Select a data feature from all data features to split the node and enter Step3.

Step 3: Generate several child nodes, judge each child node, if the stop-splitting condition is met, enter Step 4; otherwise, enter Step 2.

Step 4: Set this node as a child node, and its output result is the category with the largest proportion of this node's quantity.

From the above steps, it can be seen that there are two important problems in the decision-making process: how to select the splitting feature and when to stop splitting.

Selection of Splitting Attributes

Decision trees use greedy algorithms for splitting, meaning they split on attributes that yield the best splitting results. So, what makes a splitting result the best? The ideal situation is to find an attribute that can perfectly separate different classes, but in most cases, such a perfect split is hard to achieve at once. It is hoped that after each split, the data of child nodes is as "pure" as possible, as shown in Figs. 5.5 and 5.6

From Figs. 5.5 and 5.6, it is clear that the child nodes after splitting on Attribute 2 are purer than those after splitting on Attribute 1. After splitting on Attribute 1, the number of classes at each node remains the same, and the classification results compared to the root node are not improved at all. After splitting on Attribute 2, the number of classes at each node varies significantly, and it is highly probable that the output of the first child node is Class 1, and the output of the second child node is Class 2.

The key to splitting attributes is to find the purest state of all child node data. Decision trees use information gain or information gain ratio as the basis for selecting attributes.

Fig. 5.5 Split attribute 1

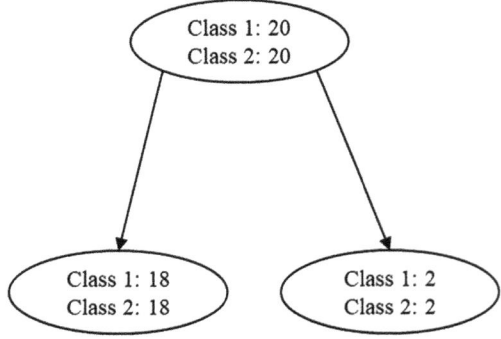

Fig. 5.6 Split attribute 2

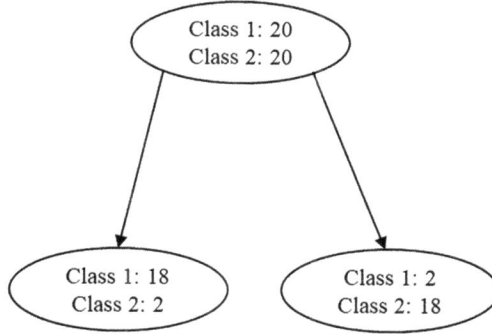

Information Gain

Information gain is used to represent the change in data complexity before and after splitting and the data complexity of splitting nodes. The formula is calculated as:

$$Info_{Gain} = Gain - \sum_{i=1}^{n} Gain;$$

Among them, *Gain* represents the complexity of nodes. The higher the *Gain*, the higher the complexity. Simply put, information gain is the sum of the data complexity before splitting minus the data complexity of child nodes. The greater the information gain, the more the data complexity decreases after splitting, and the more distinct the classification.

There are two different calculation methods for the complexity of nodes.

1. Entropy describes the degree of data chaos. The higher the entropy, the higher the degree of chaos, which means the lower the purity; conversely, the lower the entropy, the lower the degree of chaos, which means the higher the purity. The formula for calculating entropy is:

$$Entropy = -\sum_{i=1}^{n} P_i \log P_i$$

Among them, P_i represents the proportion of the number of Class i. For example, in binary classification, if the number of two classes is the same, The purity is the lowest, and the entropy equals 1; if the data of the node belong to the same class, then the purity of the node is the highest at this time, and the entropy equals 0.

2. The Gini value is calculated as follows:

$$G_{ini} = 1 - \sum_{i=1}^{n} p_i^2$$

Among them, p_i represents the proportion of the number of Class i. Still taking the above binary classification example, when the numbers of the two classes are equal, the Gini value equals 0.5; when the data of the node belong to the same class, the Gini value equals 0. The larger the Gini value, the less pure the data.

Example 5.6 Using entropy as the statistical measure of node complexity, calculate the information gain of the following examples respectively. Figure 5.7 shows the result of splitting the node by selecting Attribute 1, and Fig. 5.8 shows the result of splitting the node by selecting Attribute 2. By calculating the information gain after splitting the two attributes, select the optimal splitting attribute.

Solution The solution process for splitting attribute 1 shown in Fig. 5.7 is as follows.

$$Info1 = entropy - \sum_{i=1}^{n} entropy_i = \left(\frac{25}{25+20}\right)\log\left(\frac{25}{25+20}\right) + \left(\frac{25}{25+20}\right)\log\left(\frac{25}{25+20}\right)$$

$$\rightarrow entropy - \left[\left(\frac{19}{19+5}\right)\log\left(\frac{19}{19+5}\right) + \left(\frac{19}{19+5}\right)\log\left(\frac{19}{19+5}\right)\right]$$

Fig. 5.7 Split attribute 1

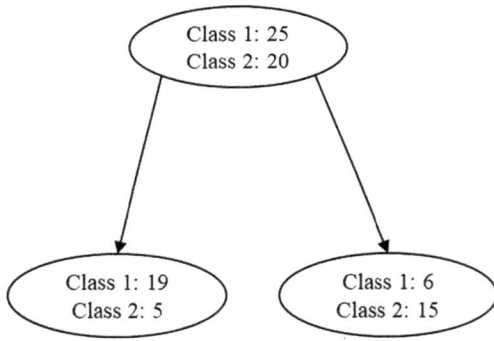

Fig. 5.8 Split attribute 2

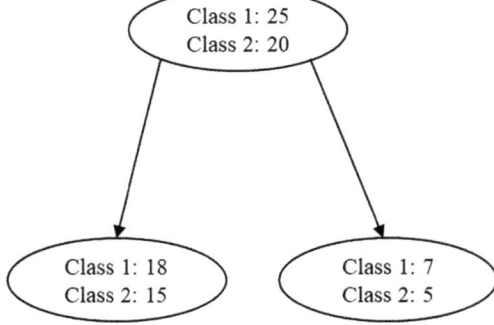

5.2 Decision Trees

$$\rightarrow entropy_1 - \left[\left(\frac{6}{15+6}\right)\log\left(\frac{6}{15+6}\right)+\left(\frac{6}{15+6}\right)\log\left(\frac{6}{15+6}\right)\right]$$

$$\rightarrow entropy_2 = 0.423$$

The solution process for splitting attribute 2 shown in Fig. 5.8 is as follows.

$$Info2 = entropy - \sum_{i=1}^{n} entropy_i = \left(\frac{25}{25+20}\right)\log\left(\frac{25}{25+20}\right)+\left(\frac{25}{25+20}\right)\log\left(\frac{25}{25+20}\right)$$

$$\rightarrow entropy - \left[\left(\frac{15}{15+18}\right)\log\left(\frac{15}{15+18}\right)+\left(\frac{15}{15+18}\right)\log\left(\frac{15}{15+18}\right)\right]$$

$$\rightarrow entropy_1 - \left[\left(\frac{5}{5+7}\right)\log\left(\frac{5}{5+7}\right)+\left(\frac{5}{5+7}\right)\log\left(\frac{5}{5+7}\right)\right]$$

$$\rightarrow entropy_2 = 0.6812$$

Since Info2 > Info1, compared with attribute 1, attribute 2 is a better splitting attribute, so attribute 1 is selected as the splitting attribute.

Information Gain Ratio

Using information gain as a condition for selecting splitting has an unavoidable drawback: it tends to select attributes with more branches for splitting. To solve this problem, the concept of information gain ratio is introduced. The information gain ratio is the information gain divided by the information entropy of splitting node data, and its calculation formula is as follows:

$$Info\_Ratio = \frac{Info\_Gain}{InstrinsicInfo}$$

Among them, $Info\_Gain$ represents information gain, and $IntrinsicInfo$ represents the information entropy of splitting node data. The calculation formula of $IntrinsicInfo$ is as follows:

$$InstrinsicInfo = -\sum_{i=1}^{n}\frac{n_i}{N}\bullet \log\left(\frac{n_i}{N}\right)$$

Among them, n represents the number of child nodes, n_i represents the data quantity of the i-th child node, and N represents the data quantity of the parent node. That is to say, $IntrinsicInfo$ is the entropy of splitting nodes. If the data chain of nodes is closer, *IntrinsicInfo* is larger. If the child nodes have a more uniform distribution, *IntrinsicInfo* is larger. As the number of child nodes increases, *IntrinsicInfo* increases, and *Info _ Ratio* decreases, which can reduce the tendency of selecting splitting attributes with more child nodes. The higher the information gain ratio, the better the splitting effect.

Example 5.7 Calculate the information gain ratio after splitting the attributes in Figs. 5.7 and 5.8.

Solution The information gain ratio of attribute 1 is calculated as follows:

$$Info\_Gain_1 = 0.423$$

$$InstrinsicInfo_1 = -\left[\left(\frac{24}{24+21}\right)\log\left(\frac{24}{24+21}\right) + \left(\frac{24}{24+21}\right)\log\left(\frac{24}{24+21}\right)\right] = 0.6909$$

$$Info\_Ratio_1 = \frac{Info\_Gain_1}{InstrinsicInfo_1} = 0.6122$$

The information gain ratio of attribute 2 is calculated as follows:

$$Info\_Gain_2 = 0.6812$$

$$InstrinsicInfo_2 = -\left[\left(\frac{33}{33+12}\right)\log\left(\frac{33}{33+12}\right) + \left(\frac{33}{33+12}\right)\log\left(\frac{33}{33+12}\right)\right] = 0.5799$$

$$Info\_Ratio_2 = \frac{Info\_Gain_2}{InstrinsicInfo_2} = 1.1747$$

Conditions for Stopping Splitting

Decision trees cannot grow indefinitely, and there are always times to stop splitting. The most extreme case is when the tree automatically stops splitting when only one data point remains, but in this case, the tree is too complex, and the prediction accuracy is not high. In general, in order to reduce the complexity of the decision tree and improve the prediction accuracy, the splitting of nodes will be terminated in advance.

The following are the general conditions for stopping the splitting of decision tree nodes:

1. Minimum number of nodes. When the data quantity of nodes is less than a specified quantity, splitting is not continued. There are mainly two reasons: one is that when the data quantity is small, splitting is prone to the influence of noise data;

5.2 Decision Trees

the second is to reduce the complexity of tree growth. Ending splitting in advance can help to reduce over-fitting to a certain extent.
2. Entropy or Gini value is less than the threshold. The size of entropy and Gini value indicates the complexity of data. When entropy or Gini value is too large, it indicates that the data purity is relatively low. When entropy or Gini value is less than a certain degree, node splitting stops.
3. The depth of the decision tree reaches the specified condition. The depth of a node can be understood as the distance between the node and the root node of the decision tree. For example, the depth of a child node of the root node is 1, because the distance between these nodes and the root node is 1. The depth of a child node is deeper than that of its parent node. The depth of the decision tree is the maximum depth of all leaf nodes. When the depth reaches the specified upper limit, splitting stops.
4. All features have been used, and splitting cannot continue. This is a condition for automatic stopping of splitting. When there are no more divisible features, the current node setting is directly set as a leaf node.

5.2.3 Pruning of Decision Trees

The basic strategies of decision tree pruning are "pre-pruning" and "post-pruning" (He and Zhang 2021).

Pre-pruning refers to making a prior assessment for each node in the process of decision tree generation. If the division of the current node cannot bring about an improvement in the generalization performance of the decision tree, then stop the division and mark the current node as a leaf node.

Post-pruning is to first generate a complete decision tree from the training set, and then traverse from the bottom up to non-leaf nodes. If replacing the child nodes corresponding to this node with leaf nodes can bring about an improvement in the generalization performance of the decision tree, then replace the child nodes with leaf nodes.

Pre-pruning

Pre-pruning makes many branches of the decision tree not "expand", which not only reduces the risk of over-fitting, but also significantly reduces the training time cost and testing time cost of the decision tree.

However, when using pre-pruning decision-making, although the current division of some branches may not improve the generalization performance, and may even lead to a temporary decline in generalization performance, the subsequent division based on it may lead to a significant improvement in performance. Pre-pruning is based on the greedy algorithm, which essentially prohibits these branches from expanding, bringing the risk of under-fitting to the pre-pruning decision tree.

In the process of tree generation, certain criteria are set to determine whether to continue growing the tree. For example, setting the maximum height (number of layers) of the decision tree to limit the growth of the tree, or setting the minimum number of samples that each node must contain. When the number of samples in the node is less than a certain value, the division is stopped. It can also be used to evaluate the goodness of division by using information gain and other metrics when constructing the tree. If the division of a node will lead to a subset with a lower-than-preset threshold, then the further division of the given subset is stopped. However, the selection of an appropriate threshold is difficult. A higher threshold may lead to over-simplified trees, and a lower threshold may lead to too few simplifications of the tree. The threshold needs to be determined based on the knowledge of the specific application field or through multiple test evaluations.

Post-pruning

Post-pruning decision trees usually retain more branches than pre-pruning decision trees. Under normal circumstances, the under-fitting risk of post-pruning decision trees is very small, and the generalization performance is often better than that of pre-pruning decision trees.

The post-pruning process is carried out after the complete decision tree is generated, and all non-leaf nodes in the tree need to be examined one by one from the bottom up, so its training time cost is much larger than that of un-pruned decision trees and pre-pruning decision trees.

First, allow the tree to grow as much as possible, and then prune the nodes of the tree by deleting the branches of the nodes to trim the tree to a smaller size. Of course, while pruning, it is also required to try to keep the accuracy of the decision tree from dropping too much. The calculations required for the post-pruning method are more than those for the pre-pruning method, but usually a more reliable tree is produced.

5.2.4 Implementation of Decision Tree Algorithms

ID3 Algorithm

The ID3 algorithm can only handle discrete-type variables, and a variable will not be used again once it has been used.

First, define the information gain after a node selects a discrete-type variable X and generates several child nodes according to its values. Use T to represent the mother node, the data sample at t, and record the sample size of the mother node as T, and the entropy as I, while the sample sizes of its individual child nodes and sample sizes are respectively X and $T_1, T_2, \ldots, T_n$. If the entropy of the mother node

5.2 Decision Trees

is $I(T)$, and the entropies of the child nodes obtained by variable X are $I(X, T_1)$, $I(X, T_2)$, ..., $I(X, T_n)$, then the definition is

$$I(X,Y) = \sum_{i=1}^{n} \frac{|T_i|}{|T|} I(X,T)$$

The information gain of this node for variable X is defined as

$$Gain(X,T) = I(T) - I(X,I)$$

Obviously, the information gain from variable X represents the information needed for identifying elements in T and the information needed for identifying elements in T after obtaining X. Information gain ranks each node for variable selection, and the variable with the largest information gain continues to construct the tree at this node. The intention of doing this is to create a tree with the smallest possible generation or to make the tree grow fastest.

Figure 5.9 is a formal form of the ID3 algorithm.

C4.5 Algorithm

The C4.5 algorithm is similar to the generation process of the ID3 algorithm decision tree. The C4.5 algorithm has improved the ID3 algorithm, using the information gain ratio selection feature. This improvement is mainly for sample characteristics.

1. The basic decision tree requires that the feature A takes values as discrete values. If A is a continuous value, if A has v values, then the test of feature A can be regarded as $v - 1$ possible tests. In fact, this process can be seen as a discretization process, but the gap between these discrete values will be relatively small. Of course, other methods can also be used, such as dividing the continuous values into segments, and then setting sub-variables.

---

function ID3 (R: the set of variables not yet used, T: the training data set at this node)
    if T is an empty set, return failure information;
    if T contains values of all the same classification variables, return a single node with that value;
    if R is empty, then return a current variable value with the maximum frequency;
    Let D ∈ R be the variable with the maximum Gain(D, T);
    Let {di} (i = 1, 2,..., m) be the values of D;
    Let {Si} (i = 1, 2,..., m) be the subsets of T corresponding to the values of D;
    Return the node labeled with D and the branches labeled with d1, d2,..., dm;
end ID3;

Note: At this time, the functions and parameters of ID3 are ID3(R - {D}, T1), ID3(R - {D}, T2),..., ID3(R - {D}, Tm).

Fig. 5.9 ID3 algorithm form code

2. Each value taken by feature A will generate a branch, and sometimes the sample size of the divided subset will be too small, and the statistical characteristics are not sufficient and continuous division cannot continue. In this case, when marking categories like this, local errors will also occur. For this situation, a group of values of A can be taken as branch conditions; or a binary decision tree can be adopted, and each branch represents a feature test situation (only whether there are two values).
3. Some samples have missing values on feature A. For this kind of null value situation, many methods can be used. For example, the values with the most occurrences of feature A in other samples can be used to fill the gaps. In some fields, the smoothing of sample internal values can be used for compensation. When the sample size is large, the samples with these missing values can also be discarded.
4. As the data set continues to decrease, the sample size of the subset will become smaller and smaller, and the constructed decision tree may have fragments, repetitions, and replications. At this time, new features can be modeled by using the original features of the sample.
5. The information gain method tends to select features with more combined values (this is determined by the definition of information entropy). For this problem, people have proposed the gain ratio method, which considers the probability of each feature value, including the K-Nearest Neighbor method, G-Statistic method, etc.

CART Algorithm

The CART algorithm can both make classifications and can also make regressions, and can only form binary trees. The branch conditions of the CART algorithm are binary classification problems.

For continuous feature situations, the branching method of the CART algorithm is to compare thresholds. If the threshold is higher than a certain value, it belongs to one category, and if it is lower than a certain value, it belongs to another category. For discrete features, the branching method is to extract sub-features. For example, for the feature of appearance, there are three levels: handsome, ugly, and medium. First, it can be divided into handsome and not-handsome, and then the not-handsome ones are further divided into ugly and medium.

The scoring function of the CART algorithm for tree picking is the result of the most classification (that is, the mode) for discrete values, and the average value for continuous values.

The loss function of the CART algorithm is the classification criterion, that is, to find the most optimized criterion. For classification trees (target variables are discrete variables), the loss function is the average of the Gini coefficients of all branch hypothesis functions. For regression trees (target variables are continuous variables), the loss function is the mean square error of all branch hypothesis functions.

For classification trees (target variables are discrete variables), the Gini coefficient is used as the splitting rule. The Gini coefficient before splitting minus the

5.2 Decision Trees

Gini coefficient after splitting, the less the reduction, the more the splitting rule is selected. For regression trees (target variables are continuous variables), the least mean square error is used as the splitting rule to generate binary trees.

Improvements of the CART Classification Tree Algorithm for Processing Continuous and Discrete Features

For the processing of CART classification tree continuous values, its idea is the same as that of the C4.5 algorithm, which is to discretize all continuous features. The only difference is that when choosing splitting points, the measurement method used by C4.5 is information gain ratio, while the CART classification tree uses the Gini coefficient.

The specific steps are as follows. For example, if there are m sample continuous features A with m values, ranked from smallest to largest as $a_1, a_2, ..., a_m$, then the CART algorithm takes the average of two adjacent sample values. A total of $m-1$ partition points are obtained, and the $i-th$ partition point is expressed as $T_i = (a_i + a_{i+1})/2$. For these $m-1$ points, calculate the Gini coefficient for each point as a binary classification point. Select the smallest Gini coefficient as the binary discrete classification point for this continuous feature. It should be noted that, whether it is ID3 or C4.5, when dealing with discrete attributes, if the current node is for continuous attributes, then this attribute may still be involved in the generation process of child nodes.

For the problem of dealing with continuous values of CART classification trees, the method adopted is to continuously perform binary discrete classification.

For ID3 or C4.5, if a certain feature A is selected to build a decision tree node, if it has three types A_1, A_2, A_3, we will create a three-branch node on the decision tree, resulting in a multi-branch tree. However, the method used by CART classification trees is different; it uses binary partition. For m sample features, CART classification trees will consider dividing A into $\{A_1\}$ and $\{A_2, A_3\}$, $\{A_2\}$ and $\{A_1, A_3\}$, $\{A_3\}$ and $\{A_1, A_2\}$ three situations, find the smallest Gini coefficient, such as $\{A_2\}$ and $\{A_1, A_3\}$, and then create a binary tree node. One node is the corresponding of A2, and the other node is the corresponding of $\{A_1, A_3\}$. At the same time, since the values of feature A are not completely divided, there is still a chance to continue to divide child nodes according to feature A, which is the same for ID3 or C4.5. In a tree generated by ID3 or C4.5, discrete features will only be involved in the generation process of a node once.

The Specific Flow of the CART Classification Tree Building Algorithm

The above introduces some differences between the CART algorithm and the C4.5 algorithm. The following introduces the specific flow of the CART classification tree building algorithm. Because it is built, it is because the CART tree algorithm also has a pruning method.

The input of the algorithm includes the training set D, Gini coefficient, and sample number threshold. The algorithm output is the decision tree T.

The algorithm starts from the root node and uses the training set to recursively build the CART tree.

1. For the current node dataset is D, if the sample number is less than the threshold or there is no feature, then return the decision tree node, and the current node stops splitting.
2. Calculate the Gini coefficient of the dataset D. If the Gini coefficient is less than the threshold, then return the decision tree node, and the current node stops splitting.
3. Calculate the Gini coefficient of each existing feature value for the dataset D. The treatment methods for discrete values and continuous values and the calculation of the Gini coefficient are described in Sect. 5.2.2. The treatment methods for missing values are the same as those described in the C4.5 algorithm.
4. Among the calculated Gini coefficients of each feature value for the dataset D, select the feature A with the smallest Gini coefficient and the corresponding threshold value α. According to this optimal feature and optimal value, divide the dataset into D_1 and D_2 parts. At the same time, establish the left and right nodes of the current node. The left node dataset is D1, and the right node dataset is D_2 (because it is a binary tree, so $D_2 = D - D_1$ here).
5. Recursively call (1) ~ (4) steps for the left and right child nodes to generate a decision tree.

CART Regression Tree Building Algorithm

The algorithms for building CART regression trees and CART classification trees are largely similar, so only the differences in the building algorithms of CART regression trees and CART classification trees are discussed here.

Firstly, the difference between regression trees and classification trees lies in the sample output. If the sample output is a discrete value, then it is a classification tree; if the sample output is a continuous value, then it is a regression tree.

Apart from the difference in concepts, there are two main differences in the building and prediction of CART regression trees and CART classification trees:

1. The processing methods of continuous values are different.
2. The ways of making predictions after the decision tree is built are different.

For the processing of continuous values, the CART classification tree uses the Gini coefficient to measure the quality of each split point of the size measure feature. This is more suitable for classification models, but for regression models, common measurement methods such as sum of squares and variance can be used. The measurement objective of the CART regression tree is, for any split feature A, corresponding to any split point s, to divide the data sets D_1 and D_2 on both sides, find the sum of variances of D_1 and D_2 to be the smallest, and at the same time, the sum of

5.2 Decision Trees

variances of D_1 and D_2 is the smallest for the corresponding feature and feature value split point. The expression is:

$$\min_{A,s}\left[\min_{c_1}\sum_{x_i\in D_1(A,s)}(y_i-c_1)^2+\min_{c_2}\sum_{x_i\in D_2(A,s)}(y_i-c_2)^2\right]$$

Where c_1 is the sample mean output value of dataset D_1; c_2 is the sample mean output value of dataset D_2.

For the way of making predictions after the decision tree is built, it was mentioned above that CART classification trees use the class with the highest probability at the leaf nodes as the prediction category for the current node. However, the regression tree output is not a category, instead, it uses the mean value of the final leaf nodes or the median value of the output prediction results.

Pruning of CART Trees

The pruning strategies of CART regression trees and CART classification trees are basically the same except for the use of mean square error when measuring losses.

Since the decision-making algorithm is very easy to over-fit the training set, resulting in poor generalization ability, to solve this problem, it is necessary to prune CART trees, which is similar to the regularization of linear regression to increase the generalization ability of decision trees. However, there are many pruning methods, so how should we choose? CART adopts the post-pruning method, that is, after the decision tree is generated, all possible pruned CART trees are generated, and then cross-validation is used to test the effectiveness of each pruning method, and the pruning strategy with the best generalization ability is selected.

In other words, the pruning algorithm of CART trees can be summarized in two steps. The first step is to generate decision trees of various pruning results from the original decision tree, and the second step is to use cross-validation to test the prediction ability of the pruned branches, and select the number of pruned branches with the best generalization ability as the final CART tree.

Firstly, consider the loss function of pruning. For any subtree T_t, its loss function is

$$C_\alpha(T_t)=C(T_t)+\alpha|T_t|$$

where α is the regularization parameter, which is similar to the regularization of linear regression. $C(T_t)$ is the predicted error of the training data, the classification tree is measured by the Gini coefficient, and the regression tree is measured by the mean square error. $|T_t|$ is the number of leaf nodes of the subtree T_t.

When $\alpha = 0$, there is no regularization, and the original generated CART tree is the optimal subtree. When $\alpha = \infty$, the regularization intensity reaches the maximum, and at this time, the single-node tree composed of the root nodes of the originally generated CART tree is the optimal subtree. Of course, these are two extreme cases. Generally speaking, the larger α is, the more severe the pruning is, and the generated optimal subtree is smaller compared to the original generated decision tree. For a fixed α, there must exist a single subtree that minimizes the loss function $C_\alpha(T_t)$.

After understanding the loss function of pruning, let's look at the idea of pruning again. For any leaf-node subtree T_t at node t, if there is no pruning, its loss is:

$$C_\alpha(T_t) = C(T_t) + \alpha |T_t|$$

If it is pruned, only the root node is retained, and the loss is:

$$C_\alpha(T_t) = C(T_t) + \alpha$$

When $\alpha = 0$ or α is very small:

$$C_\alpha(T_t) < C_\alpha(T)$$

When α increases to a certain extent:

$$C_\alpha(T_t) = C_\alpha(T)$$

When α continues to increase, the inequality is reversed, that is, if the following is satisfied:

$$\alpha = \frac{C(T) - C(T_t)}{|T_t| - 1}$$

T_t and T have the same loss function, but T has fewer nodes, so the subtree T_t can be pruned, that is, all the child nodes of this node are cut off, becoming a leaf-node T.

Finally, look at the cross-validation strategy of the CART tree. Through the above steps, the threshold α of whether each subtree is pruned can be calculated. If the values of all nodes that are not pruned are calculated, and then the best subtree after pruning corresponding to different α values is cross-validated. In this way, the best α can be selected, and with this α, the corresponding best subtree can be used as the final result.

Pruning Algorithm of CART Tree

The input is the original decision tree T obtained by the CART tree building algorithm, and the output is the optimal decision subtree T_α.

The algorithm steps are as follows:

1. Initialize $\alpha_{min} = \infty$, and the optimal subtree set $\omega = \{T\}$.

5.2 Decision Trees

2. Starting from the leaf nodes, calculate the training error loss function $C_\alpha(T_t)$ (the mean square error for regression trees and the Gini coefficient for classification trees), the number of leaf nodes $|T_t|$, and the regularization threshold value for each internal node t from bottom to top:

$$\alpha = \min\left\{\frac{C(T)-C(T_t)}{|T_t|-1}, \alpha_{\min}\right\}$$

Update $\alpha_{\min} = \alpha$.

3. Obtain the set M of α values for all nodes.
4. Select the maximum value α_k from M and visit the internal nodes of the subtree t from top to bottom. If $\frac{C(T)-C(T_t)}{|T_t|-1} \leq \alpha_k$, perform pruning and determine the value of the leaf node t. If it is a classification tree, it is the class with the highest probability; if it is a regression tree, it is the mean of all sample outputs. In this way, the optimal subtree T_k corresponding to α_k is obtained.
The optimal subtree set $= \omega \cup T_k$, $M = M - \{\alpha_k\}$.
5. If M is not empty, return to step (4); otherwise, obtain all the optimal subtree sets ω.
6. Use cross-validation to select the optimal subtree T_α from ω.

Steps of the Decision Tree Generation Algorithm

The input is the training data set S and the attribute set (including class label attributes); the output is the decision tree T. The specific steps are as follows:
   Create a node R.

1. If all data samples in the sample set S are in the same class C, then R is a leaf node, and label the leaf node with class C.
2. If there are no remaining attributes, then R is a leaf node, and label the leaf node with the majority class label in the sample.
3. Calculate the attribute selection measure (such as information gain or Gini coefficient, etc.) for each attribute A, and determine the splitting attribute A_a.
4. Use the attribute A_a as the splitting attribute of the node R.
5. For each split value α_i of the splitting attribute, use different methods to handle the cases where the attribute value is discrete or continuous. When the value is discrete, the method is straightforward. When the value is continuous, the following steps are needed.
6. Split the node R with a condition $S_t = \alpha_i$.
7. Let S_1 be the sample subset of the sample set S where $S_{alt} = \alpha_i$.

8. If the sample set *S* is empty, add a leaf node, and label it with the most common class in the sample set *S*; otherwise, add a node returned by the statement *building _ tree* (S, removing the attribute set of the splitting attribute).

5.2.5 Relevant Applications of Decision Trees and MATLAB Calculation Examples

Application Example 1: Psychological Activity Judgment Based on Information

The decision tree algorithm is constructed based on information gain, and information gain can be calculated from the entropy of the training set.

Example 5.8

$$data = \begin{bmatrix} good\ mood & good\ weather & go\ out \\ good\ mood & bad\ weather & go\ out \\ bad\ mood & good\ weather & go\ out \\ bad\ mood & bad\ weather & don't\ go\ out \end{bmatrix}$$

Find its information gain.

Solution The first two columns are classification attributes, and the last column is classification. The entropy of classification can be calculated as follows:

$$go\ out = 3, don't\ go\ out = 1, total\ number = 4$$

The first column of attributes has two types: good mood and bad mood.

$$good\ mood, go\ out = 2, don't\ go\ out = 0, total\ number = 2$$

The information entropy of being in a good mood
$$= -(2/2)\log 2(2/2) + (0/2)\log 2(0/2)$$

Similarly,

The information entropy of being in a bad mood
$$= -(1/2)\log 2(1/2) - (1/2)\log 2(1/2)$$

5.2 Decision Trees

*The information gain of mood = classification information entropy
− probability of being in a good mood × information entropy
of being in a good mood − probability of being in a bad mood
× information entropy of being in a bad mood*

Thus, the information entropy corresponding to each attribute can be obtained, and the attribute with the largest information entropy is the optimal partitioning attribute.

Example 5.9 Based on Example 5.4, add the optimal partitioning attribute as mood, as shown in Fig. 5.10, and find its decision tree.

Solution Distinguish whether the classification under each specific situation of the mood attribute is all the same. If they are the same, mark the node as this category. In the case of a good mood, regardless of the weather, the result is going out, as shown in Fig. 5.11.

In the case of a bad mood, there are different classification results. Continue to calculate the information gain of other attributes under the condition of a bad mood, and use the attribute with the largest information gain as the branch node. Here, there is only the weather attribute, so this node is weather. The weather attribute has two situations, as shown in Fig. 5.12.

In the case of a bad mood and good weather, if the classification is all the same, change the node label to this category. In the case of a bad mood and good weather, the result is going out; in the case of a bad mood and bad weather, the result is not going out, as shown in Fig. 5.13.

Example 5.10 For the attributes under the branch nodes, there may be no data. Assume the training set becomes

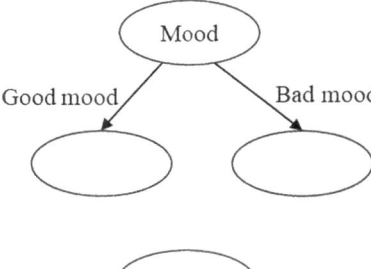

Fig. 5.10 Optimal attribute partition

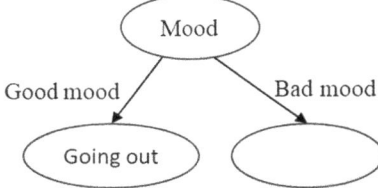

Fig. 5.11 Results of optimal attribute division

Fig. 5.12 Optimal attribute partition-weather

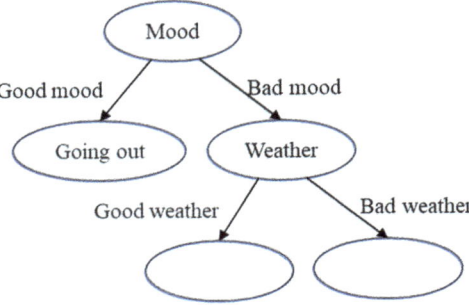

Fig. 5.13 Optimal attribute partition-weather results

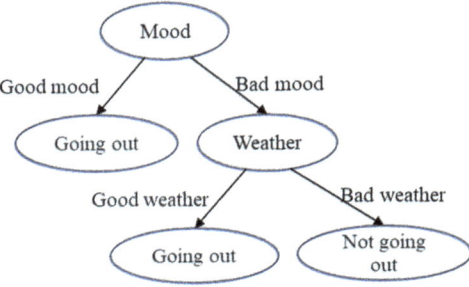

$$\begin{bmatrix} good\ mood & sunny & go\ out \\ good\ mood & cloudy & go\ out \\ good\ mood & rainy & go\ out \\ good\ mood & foggy & go\ out \\ bad\ mood & sunny & go\ out \\ bad\ mood & rainy, & don't\ go\ out \\ bad\ mood & cloudy & don't\ go\ out \end{bmatrix}$$

Find its decision tree.

Solution In the case of a bad mood, there is no foggy weather. How to judge whether to go out in foggy weather? We can use the most common classification of this sample as the classification. In the case of a bad mood, *go out* = 1, *don't go out* = 2, so *don't go out* is taken as the classification result of foggy weather, as shown in Fig. 5.14.

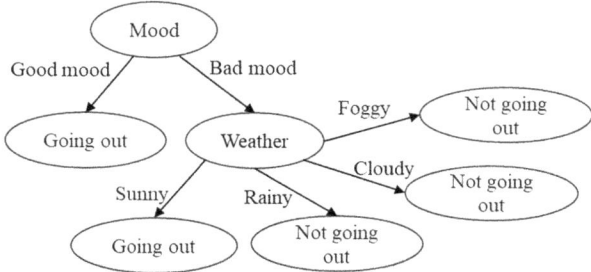

Fig. 5.14 Optimal attribute partition results

At this point, the partitioning of all attributes is completed, and recursive partitioning is ended, resulting in a very simple decision tree. Its MATLAB code implementation is as follows:

```
function [node] = createTree(data, feature)
    type = mostType(data);
    [m,n] = size(data);
    node = struct('value','null','name','null','type','null','children',[]);
    temp_type = data(1,n);
    temp_b = true;
    for i=1:m
        if temp_type~=data(i,n)
            temp_b = false;
        end
    end
    if temp_b==true
        node.value = data(1,n);
        return;
    end
    if sum(feature) == 0
        node.value = type;
        return;
    end
    feature_bestColumn = bestFeature(data);
    best_feature = getData(:,feature_bestColumn);
    best_distinct = unique(best_feature);
    best_num = length(best_distinct);
    best_proc = zeros(best_num,2);
    best_proc(:,1) = best_distinct(:,1);
    for i = 1:best_num
        Dv = [];
Dv_index = 1;
```

```
bach_node = struct('value', 'null', 'name', 'null', 'type',
'null', 'children', []);
for j = 1:m
    if best_proc(i, 1) == data(j, feature_bestColumn)
        Dv(Dv_index, :) = data(j, :);
        Dv_index = Dv_index + 1;
    end
end
if length(Dv) == 0
    bach_node.value = type;
    bach_node.type = best_proc(i, 1);
    bach_node.name = feature_bestColumn;
    node.children(i) = bach_node;
    return;
else
    feature(feature_bestColumn) = 0;
    bach_node = createTree(Dv, feature);
    bach_node.type = best_proc(i, 1);
    bach_node.name = feature_bestColumn;
    node.children(i) = bach_node;
end
end

function [column] = bestFeature(data)
    [m, n] = size(data);
    featureSize = n - 1;
    gain_proc = zeros(featureSize, 2);
    entropy = getEntropy(data);
    for i = 1:featureSize
        gain_proc(i, 1) = i;
        gain_proc(i, 2) = getGain(entropy, data, i);
    end
    for i = 1:featureSize
        if gain_proc(i, 2) == max(gain_proc(:, 2))
            column = i;
            break;
        end
    end
end

function [res] = mostType(data)
    [m, n] = size(data);
    res_distinct = unique(data(:, n));
    res_proc = zeros(length(res_distinct), 2);
    res_proc(:, 1) = res_distinct(:, 1);
```

5.2 Decision Trees

```
    for i = 1:length(res_distinct)
        for j = 1:m
            if res_proc(i, 1) == data(j, n)
                res_proc(i, 2) = res_proc(i, 2) + 1;
            end
        end
    end
    end
    for i = 1:length(res_distinct)
        if res_proc(i, 2) == max(res_proc(:, 2))
            res = res_proc(i, 1);
            break;
        end
    end
    end

    function [entropy] = getEntropy(data)
        entropy = 0;
        [m, n] = size(data);
        label = data(:, n);
        label_distinct = unique(label);
        label_num = length(label_distinct);
        proc = zeros(label_num, 2);
        proc(:, 1) = label_distinct(:, 1);
        for i = 1:label_num
            for j = 1:m
                if proc(i, 1) == data(j, n)
                    proc(i, 2) = proc(i, 2) + 1;
                end
            end
        end
        for i = 1:label_num
            if proc(i, 2) ~= 0
                entropy = entropy - proc(i, 2) / m *
log2(proc(i, 2) / m);
            end
        end
    end

    function [gain] = getGain(entropy, data, column)
        [m, n] = size(data);
        feature = data(:, column);
        feature_distinct = unique(feature);
        feature_num = length(feature_distinct);
        feature_proc = zeros(feature_num, 2);
```

```
            feature_proc(:, 1) = feature_distinct(:, 1);
            f_entropy = 0;
            for i = 1:feature_num
                feature_data = [];
                feature_proc(i, 2) = 0;
                feature_row = 1;
                for j = 1:m
                    if feature_proc(i, 1) == data(j, column)
                        feature_proc(i, 2) = feature_proc(i, 2) + 1;
                        feature_data(feature_row, :) = data(j, :);
                        feature_row = feature_row + 1;
                    end
                end
                f_entropy = f_entropy + feature_proc(i, 2) / m *
getEntropy(feature_data);
            end
            gain = entropy - f_entropy;
        end
end
gain = entropy - f_entropy;
```

Application Example 2: Vehicle Feature Evaluation Quality

Example 5.11 Given a data set including vehicle purchase price, maintenance cost, number of doors, number of passengers, power performance, and safety performance, classify vehicles to determine the quality of a vehicle. The vehicle quality is divided into four types: non-compliant, compliant, good, and excellent. The data set format is shown in Fig. 5.15, where each value can be regarded as a string.

Solution Consider the six attributes in the data set concentration, and their value ranges are as follows:

1. Purchase price: The value range is vhigh, high, med, low, representing very high, high, medium, and low, respectively.
2. Maintenance cost: The value range is vhigh, high, med, low, representing very high, high, medium, and low, respectively.
3. Number of doors: The value range is 2, 3, 4, 5, 5more, etc.
4. Number of passengers: The value range is 2, 4, more, etc.

Fig. 5.15 The form of data set

med,low,5more,more,big,low,unacc
med,low,5more,more,big,med,good
med,low,5more,more,big,high,vgood
low,vhigh,2,2,small,low,unacc
low,vhigh,2,2,small,med,unacc
low,vhigh,2,2,small,high,unacc
low,vhigh,2,2,med,low,unacc

5.2 Decision Trees

5. Power performance: The value range is small, med, big, representing small, medium, and large, respectively.
6. Safety performance: The value range is low, med, high, representing low, medium, and high, respectively.

The classification results, that is, the vehicle quality value range is unacc, acc, good, vgood, representing non-compliant, compliant, good, and excellent, respectively.

Considering that each row has string attributes, it is necessary to assume that all features are strings, and build classifiers on this basis.

First, convert all strings in the data set concentration into numbers for easy classification later. Since the downloaded data set is in .data format and MATLAB cannot read it directly, it has been converted to .xlsx format, and vhigh, high, med, low are replaced with 4, 3, 2, 1, respectively, small, med, big are replaced with 1, 2, 3, respectively, low, med, high are replaced with 1, 2, 3, respectively, and unacc, acc, good, vgood are replaced with 1, 2, 3, 4, respectively.

There are a total of 1728 groups in the data. Randomly select 1500 groups as the training set, and the remaining 228 groups are used as the test set. Use the training set to build a decision tree, and then use the model for prediction. Calculate the prediction accuracy of the decision tree for various vehicle condition predictions and the prediction accuracy of the entire test set according to the results of the decision tree, and then prune the decision tree.

The MATLAB code implementation of the classification is as follows:

```
% Clear all variables.
clear all;
% Close all figure windows.
close all;
% Load the data.
load car;
% Generate a random permutation of integers from 1 to 1728.
a = randperm(1728);
% Training set
% Extract the first 1500 rows and the first 6 columns of the data
as training data.
Train_Data = data(a(1:1500),1:6);
% Extract the first 1500 rows and the 7th column of the data as
training labels.
Train_Label = data(a(1:1500),7);
% Test set
% Extract rows from 1501 to 1728 and the first 6 columns of the
data as test data.
Test_Data = data(a(1501:1728),1:6);
% Extract rows from 1501 to 1728 and the 7th column of the data
as test labels.
Test_Label = data(a(1501:1728),7);
```

```
% Create a decision tree classifier.
Tree = ClassificationTree.fit(Train_Data,Train_Label);
% View the decision tree.
view(Tree);
% View the decision tree in graph mode.
view(Tree,'mode','graph');
% Predict the classification.
Tree_pre = predict(Tree,Test_Data);
% Analyze the results.
% Count the number of training samples with label 1.
count_train_1_length = length(find(Train_Label == 1));
% Count the number of training samples with label 2.
count_train_2_length = length(find(Train_Label == 2));
% Count the number of training samples with label 3.
count_train_3_length = length(find(Train_Label == 3));
% Count the number of training samples with label 4.
count_train_4_length = length(find(Train_Label == 4));
% Calculate the proportion of training samples with label 1.
rate_train_1 = count_train_1_length / 1500;
% Calculate the proportion of training samples with label 2.
rate_train_2 = count_train_2_length / 1500;
% Calculate the proportion of training samples with label 3.
rate_train_3 = count_train_3_length / 1500;
% Calculate the proportion of training samples with label 4.
rate_train_4 = count_train_4_length / 1500;
% Count the number of test samples with label 1.
total_1 = length(find(Test_Label == 1));
% Count the number of test samples with label 2.
total_2 = length(find(Test_Label == 2));
% Count the number of test samples with label 3.
total_3 = length(find(Test_Label == 3));
% Count the number of test samples with label 4.
total_4 = length(find(Test_Label == 4));
% Count the number of test samples predicted as label 1.
count_test_1 = length(find(Test_Label == 1));
% Count the number of test samples predicted as label 2.
count_test_2 = length(find(Test_Label == 2));
% Count the number of test samples predicted as label 3.
count_test_3 = length(find(Test_Label == 3));
% Count the number of test samples predicted as label 4.
count_test_4 = length(find(Test_Label == 4));
% Count the number of test samples correctly predicted as
label 1.
count_right_1 = length(find(Tree_pre == 1 & Test_Label == 1));
```

5.2 Decision Trees

```
% Count the number of test samples correctly predicted as
label 2.
count_right_2 = length(find(Tree_pre == 2 & Test_Label == 2));
% Count the number of test samples correctly predicted as
label 3.
count_right_3 = length(find(Tree_pre == 3 & Test_Label == 3));
% The number of correct predictions of excellent vehicle quality
in the test set
count_right_4 = length(find(Tree_pre == 4 & Test_Label == 4));
% The number of correct predictions of good vehicle quality in
the test set
rate_right = (count_right_1 + count_right_2 + count_right_3 +
count_right_4)/228;
% Display some results
disp('Total number of vehicles: 1728');
disp(['Non - compliant: ',num2str(total_1)]);
disp(['Compliant: ',num2str(total_2)]);
disp(['Good: ',num2str(total_3)]);
disp(['Excellent: ',num2str(total_4)]) ;
disp('Number of vehicles in the training set: 1500');
disp(['Non - compliant: ',num2str(count_train_1_length)]);
disp(['Compliant: ',num2str(count_train_2_length)]);
disp(['Good: ',num2str(count_train_3_length)]);
disp(['Excellent: ',num2str(count_train_4_length)]);
disp('Number of vehicles in the test set: 228');
% Computational intelligence methods
disp('Non - compliant: ',num2str(count_test_1));
disp('Compliant: ',num2str(count_test_2));
disp('Good: ',num2str(count_test_3));
disp('Excellent: ',num2str(count_test_4));
disp('Decision tree judgment results: );
disp(['Non - compliance accuracy rate: ',sprintf('%2.2f%%',count_
right_1/count_test_1*100)]);
disp(['Compliance accuracy rate: ',sprintf('%2.2f%%',count_
right_2/count_test_2*100)]);
disp(['Good accuracy rate: ',sprintf('%2.2f%%',count_right_3/
count_test_3*100)]);
disp(['Excellent accuracy rate: ',sprintf('%2.2f%%',count_
right_4/count_test_4*100)]);
disp(['Total
accuracy rate: ',sprintf('%2.2f%%',rate_right*100)]);
% Resampling error and cross - validation error of the decision
tree before optimization
resubError = resubLoss(Tree);
lossDefault = kfoldLoss(crossval(Tree));
```

```
disp('Resampling error of the decision tree before pruning:');
disp(num2str(resubError));
disp('Cross - validation error of the decision tree before
pruning:');
disp(num2str(lossDefault));
% Pruning
[~,~,~,bestlevel] = cvloss(Tree,'subtrees','all','treesiz
e','min');
cptree = prune(Tree,'Level',bestlevel);
view(cptree,'mode','graph');
% Resampling error and cross - validation error of the decision
tree after pruning
resubPrune = resubLoss(cptree);
lossPrune = kfoldLoss(crossval(cptree));
disp('Resampling error of the decision tree after pruning:');
disp(num2str(resubPrune));
disp('Cross - validation error of the decision tree after
pruning:');
disp(num2str(lossPrune));
```

The classification results are as follows:

1. The total number of vehicles is 1728, with 1210 non - compliant, 384 compliant, 69 good, and 65 excellent.
2. The number of vehicles in the training set is 1500, with 1046 non-compliant, 338 compliant, 56 good, and 60 excellent.
3. The number of vehicles in the test set is 228, with 164 non-compliant, 46 compliant, 13 good, and 5 excellent.

The decision tree judgment results are as follows:

1. The non-compliance accuracy rate is 97.56%; the compliance accuracy rate is 95.65%; the good accuracy rate is 84.62%; the excellent accuracy rate is 100.00%; and the total accuracy rate is 96.49%.
2. The resampling error of the decision tree before pruning is 0.026.
3. The cross-validation error of the decision tree before pruning is 0.048667.
4. The resampling error of the decision tree after pruning is 0.026667.
5. The cross-validation error of the decision tree after pruning is 0.026667.

The MATLAB code implementation results are shown in Fig. 5.16.

Application Example 3: Malignant Breast Tumor Judgment

The specific MATLAB code implementation is as follows:

5.2 Decision Trees

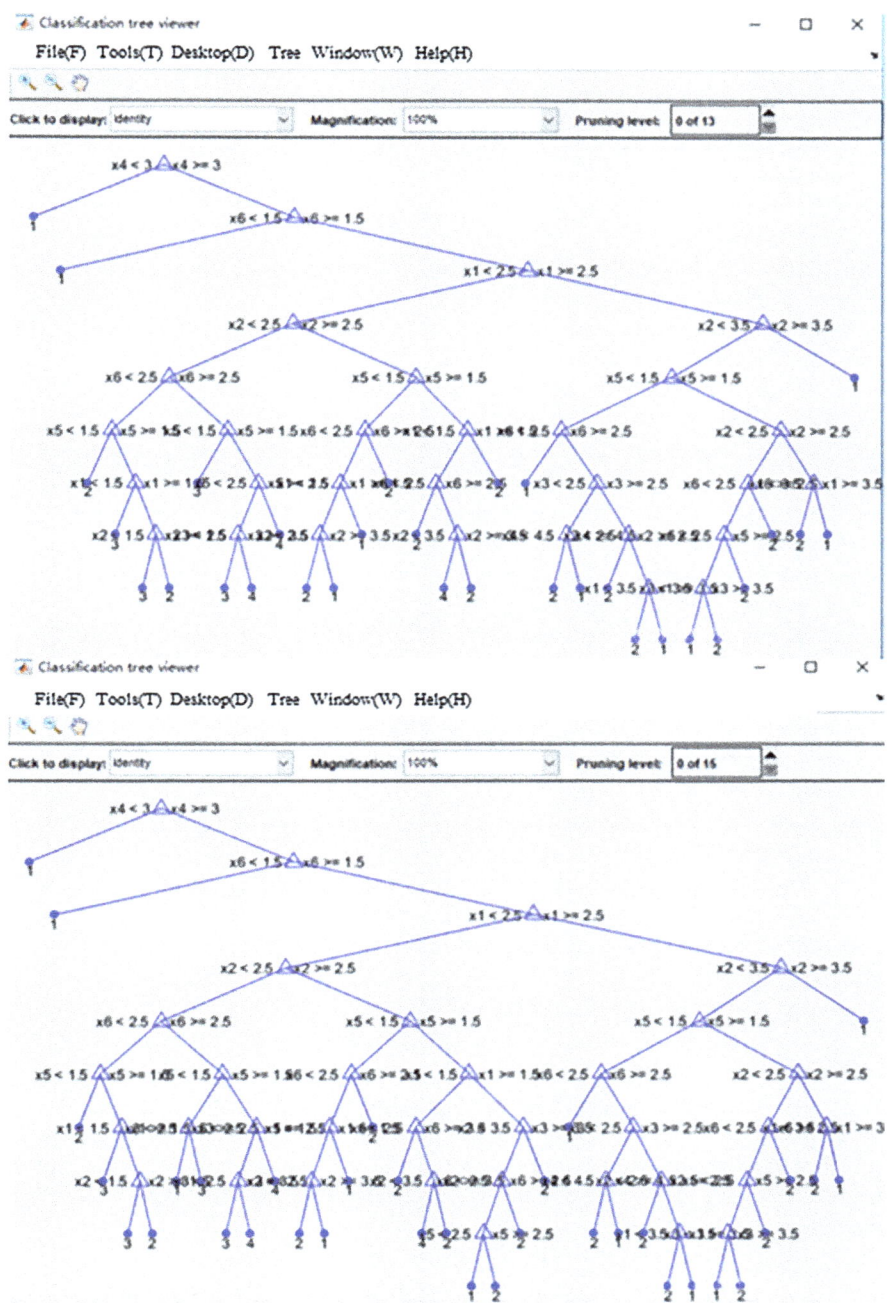

Fig. 5.16 The result of MATLAB code implementation for vehicle feature evaluation

```matlab
% Clear all variables
  clear all;
  % Clear the command window
  clc;
  % Turn off warnings
  warning off;
  % Load the data.mat file
  load data.mat;
  % Generate a random permutation of integers from 1 to 569
  a = randperm(569);
  % Select the first 500 rows of data as the training set
  Train = data(a(1:500),:);
  % Select the remaining rows of data as the test set
  Test = data(a(501:end),:);
  % Select columns 3 to end of the training set as features
  P_train = Train(:,3:end);
  % Select column 2 of the training set as labels
  T_train = Train(:,2);
  % Select columns 3 to end of the test set as features
  P_test = Test(:,3:end);
  % Select column 2 of the test set as labels
  T_test = Test(:,2);
  % Fit a classification tree using the training data
  ctree = ClassificationTree.fit(P_train,T_train);
  % View the classification tree (text format)
  view(ctree);
  % View the classification tree (graph format)
  view(ctree,'mode','graph');
  % Predict the test set using the classification tree
  T_sim = predict(ctree,P_test);
  % Count the number of benign cases (label = 1) in the
training set
  count_B = length(find(T_train == 1));
  % Count the number of malignant cases (label = 2) in the
training set
  count_M = length(find(T_train == 2));
  % Calculate the proportion of benign cases in the training set
  rate_B = count_B / 500;
  % Calculate the proportion of malignant cases in the
training set
  rate_M = count_M / 500;
  % Count the total number of benign cases in the dataset
  total_B = length(find(data(:,2) == 1));
  % Count the total number of malignant cases in the dataset
  total_M = length(find(data(:,2) == 2));
```

5.2 Decision Trees

```
% Count the number of benign cases in the test set
number_B = length(find(T_test == 1)) ;
% Count the number of malignant cases in the test set
number_M = length(find(T_test == 2)) ;
% Count the number of correctly predicted benign cases in the test set
number_B_sim = length(find(T_sim == 1 & T_test == 1));
% Count the number of correctly predicted malignant cases in the test set
number_M_sim = length(find(T_sim == 2 & T_test == 2));
% Display the total number of cases
disp('The total number of cases: num2str(569)...');
% Display the total number of benign cases
disp(['Benign: ',num2str(total_B), '...']);
% Display the total number of malignant cases
disp(['Malignant: ',num2str(total_M), '...']);
% Display the total number of cases in the training set
disp('The total number of cases in the training set: num2str(500)...');
% Display the number of benign cases in the training set
disp(['Benign: ',num2str(count_B), '...']);
% Display the number of malignant cases in the training set
disp(['Malignant: ',num2str(count_M), '...']);
% Display the total number of cases in the test set
disp('The total number of cases in the test set: num2str(69)...');
% Display the number of benign cases in the test set
disp(['Benign: ',num2str(number_B), '...']);
% Display the number of malignant cases in the test set
disp(['Malignant: ',num2str(number_M), '...']);
% Display the number of correctly diagnosed benign breast tumors
disp('The number of correctly diagnosed benign breast tumors: num2str(number_B_sim)...');
% Display the number of misdiagnosed benign cases
disp(['Misdiagnosed benign: ',num2str(number_B - number_B_sim), '...']);
% Display the diagnosis accuracy rate for benign cases (p1)
disp(['The diagnosis accuracy rate p1 = ',num2str(number_B_sim/number_B * 100), '%...']);
% Display the number of correctly diagnosed malignant breast tumors
disp('The number of correctly diagnosed malignant breast tumors: num2str(number_M_sim)...');
% Display the number of misdiagnosed malignant cases
```

```
    disp(['Misdiagnosed malignant: ',num2str(number_M - number_M_
sim), '...']);
    % Display the diagnosis accuracy rate for malignant cases (p2)
    disp(['The diagnosis accuracy rate p2 = ',num2str(number_M_
sim/number_M * 100), '%...']);
    % Find the best pruning level
    [~,~,~,bestlevel] = cvloss(ctree,'subtrees','all','treesiz
e','min');
    % Prune the classification tree
    cptree = prune(ctree,'Level',bestlevel);
    % View the pruned classification tree (graph format)
    view(cptree,'mode','graph');
```

The results of MATLAB code implementation are shown in Fig. 5.17.

Application Example 4: The Relationship Between Store Sales and Weather

Example 5.12

Build a decision—tree model for the following dataset.

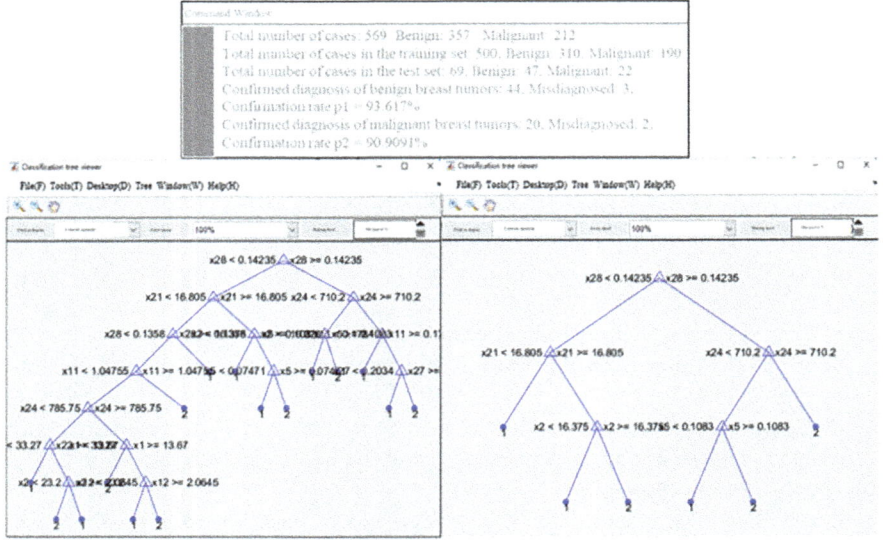

Fig. 5.17 The result of MATLAB code implementation for breast tumor classification

5.2 Decision Trees

Serial number	Weather	Is it weekend?	Is there a promotion?	Sales volume
1	Bad	Yes	Yes	High
2	Bad	Yes	Yes	High
3	Bad	Yes	Yes	High
4	Bad	Yes	Yes	High
5	Bad	Yes	Yes	High
6	Bad	Yes	Yes	High
7	Good	Yes	Yes	High
8	Good	Yes	Yes	High
9	Good	Yes	Yes	High
10	Good	Yes	Yes	High
11	Good	Yes	Yes	High
12	Good	Yes	Yes	High
13	Good	Yes	Yes	High
14	Good	No	Yes	Low
15	Good	Yes	Yes	High
16	Good	No	Yes	High
17	Good	No	Yes	High
18	Good	No	Yes	High
19	Bad	No	Yes	Low
20	Bad	No	Yes	Low
21	Bad	No	Yes	Low
22	Bad	No	Yes	Low
23	Bad	No	Yes	Low
24	Bad	Yes	Yes	High
25	Bad	Yes	Yes	High
26	Bad	Yes	Yes	High
27	Bad	Yes	Yes	High
28	Bad	Yes	Yes	High
29	Bad	Yes	Yes	High
30	Bad	Yes	Yes	High
31	Bad	Yes	Yes	High
32	Bad	Yes	Yes	High
33	Good	Yes	Yes	High
34	Good	Yes	Yes	High

Solution The specific steps of constructing a decision-tree model using the ID3 algorithm are as follows.

1. Calculate the total information entropy. Among the data, the total number of records is 34, the number of data with a sales volume of "high" is 18, and the number of data with a sales volume of "low" is 16.

$$I(18,16) = -\frac{18}{34}\log_2\frac{18}{34} - \frac{16}{34}\log_2\frac{16}{34} = 0.997503$$

2. According to

$$I(s_1, s_2, \cdots, s_m) = -\sum_{i=1}^{m} P_i \log_2(P_i)$$

$$E(A) = \sum_{j=1}^{k} \frac{s_{1j} + s_{2j} + \cdots + s_{mj}}{s} I(s_{1j}, s_{2j}, \cdots, s_{mj})$$

Calculate the information entropy of each test attribute. For the weather attribute, the attribute values are "good" and "bad". When the weather is "good", the number of records with a sales volume of "high" is 11, and the number of records with a sales volume of "low" is 6, which can be expressed as (11, 6); when the weather is "bad", the number of records with a sales volume of "high" is 7, and the number of records with a sales volume of "low" is 10, which can be expressed as (7, 10). Then the information entropy calculation process of the weather attribute is as follows:

$$I(11,6) = -\frac{11}{17}\log_2\frac{11}{17} - \frac{6}{17}\log_2\frac{6}{17} = 0.936667$$

$$I(7,10) = -\frac{7}{17}\log_2\frac{7}{17} - \frac{10}{17}\log_2\frac{10}{17} = 0.977418$$

$$E(weather) = \frac{17}{34}I(11,6) + \frac{17}{34}I(7,10) = 0.957403$$

For the weekend attribute, its values are "yes" and "no". Among them, when the weekend attribute is "yes", the number of records with high sales volume is 11, and the number of records with low sales volume is 3, which can be expressed as (11, 3); when the weekend attribute is "no", the number of records with high sales volume is 7, and the number of records with low sales volume is 13, which can be expressed as (7, 13). Then the information entropy calculation process of the weekend attribute is as follows:

$$I(11,3) = -\frac{11}{14}\log_2\frac{11}{14} - \frac{3}{14}\log_2\frac{3}{14} = 0.749595$$

$$I(7,13) = -\frac{7}{20}\log_2\frac{7}{20} - \frac{13}{20}\log_2\frac{13}{20} = 0.934068$$

$$E(weekend) = \frac{14}{34}I(11,3) + \frac{20}{34}I(7,13) = 0.858109$$

For the promotion attribute, its values are "yes" and "no". Among them, when the promotion attribute is "yes", the number of records with high sales volume is 15, and the number of records with low sales volume is 7, which can be expressed as (15, 7); when the promotion attribute is "no", the number of records with high sales volume is 3, and the number of records with low sales volume is 9, which can be expressed as (3, 9). Then the information entropy calculation process of the promotion attribute is as follows:

$$I(15,7) = -\frac{15}{22}\log_2\frac{15}{22} - \frac{7}{22}\log_2\frac{7}{22} = 0.902393$$

$$I(3,9) = -\frac{3}{12}\log_2\frac{3}{12} - \frac{9}{12}\log_2\frac{9}{12} = 0.811278$$

$$E(promotion) = \frac{22}{34}I(15,7) + \frac{12}{34}I(3,9) = 0.870235$$

According to:

$$Gain(A) = I(s_1, s_2, \cdots, s_m) - E(A)$$

Calculate the information gain of weather, weekend, and promotion attributes:

$$Gain(weather) = I(18,16) - E(weather) = 0.997503 - 0.957043 = 0.04046$$

$$Gain(weekend) = I(18,16) - E(weekend) = 0.997503 - 0.858109 = 0.139394$$

$$Gain(promotion) = I(18,16) - E(promotion) = 0.997503 - 0.870235 = 0.127268$$

3. According to the calculation results, it can be seen that the information gain of the weekend attribute is the largest, and its two attribute values "yes" and "no" are used as the root nodes of the decision tree. Then, following the above steps, continue to divide the two branches of this root node. For each branch node,

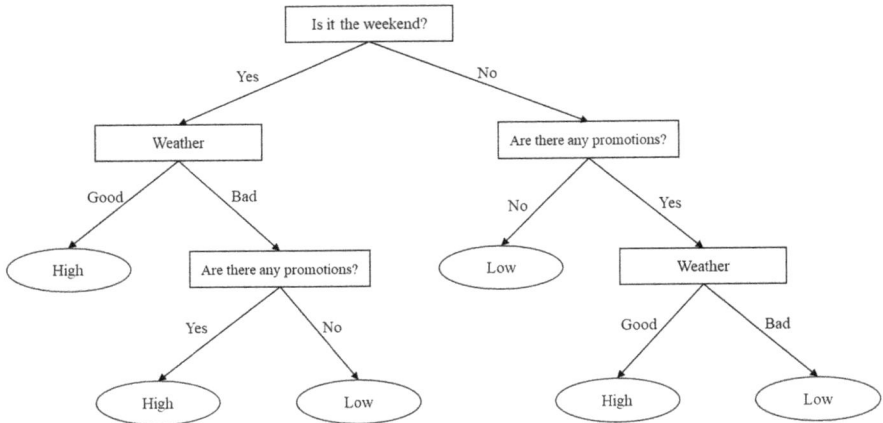

Fig. 5.18 Decision tree model

continue to calculate the information gain, repeat this process until there are no new branch nodes, and finally form a decision tree. The generated decision tree model is shown in Fig. 5.18.

According to the decision tree model, the classification results are as follows: if the weekend attribute is "yes", the weather is "good", then the sales volume is "high"; if the weekend attribute is "yes", the weather is "bad", the promotion attribute is "yes", then the sales volume is "high"; if the weekend attribute is "yes", the weather is "bad", the promotion attribute is "no", then the sales volume is "low"; if the weekend attribute is "no", the promotion attribute is "no", then the sales volume is "low"; if the weekend attribute is "no", the promotion attribute is "yes", the weather is "good", then the sales volume is "high"; if the weekend attribute is "no", the promotion attribute is "yes", the weather is "bad", then the sales volume is "low".

The MATLAB implementation code of the decision tree model is as follows:

```
clear ;
function [ matrix,attributes,activeAttributes ]   = id3_
preprocess(  )
attributes=txt(1,2:end);
activeAttributes = ones(1,length(attributes)-1);
data = txt(2:end,2:end);
[rows,cols] = size(data);
matrix = zeros(rows,cols);
for j=1:cols
    matrix(:,j) = cellfun(@trans2onezero,data(:,j));
end
end
```

5.2 Decision Trees

```
function flag = trans2onezero(data)
    if strcmp(data,'bad') ||strcmp(data,'否')...
        ||strcmp(data,'low')
        flag =0;
        return ;
    end
    flag =1;
end
function [ tree ] = id3( examples, attributes, activeAttributes )
if (isempty(examples));
    error(' Data must be provided!');
end
numberAttributes = length(activeAttributes);
numberExamples = length(examples(:,1));
tree = struct('value', 'null', 'left', 'null', 'right', 'null');
lastColumnSum = sum(examples(:, numberAttributes + 1));
if (lastColumnSum == numberExamples);
    tree.value = 'true';
    return
end
if (lastColumnSum == 0);
    tree.value = 'false';
    return
end
if (sum(activeAttributes) == 0);
    if (lastColumnSum >= numberExamples / 2);
        tree.value = 'true';
    else
        tree.value = 'false';
    end
    return
end
p1 = lastColumnSum / numberExamples;
if (p1 == 0);
    p1_eq = 0;
else
    p1_eq = -1*p1*log2(p1);
end
p0 = (numberExamples - lastColumnSum) / numberExamples;
if (p0 == 0);
    p0_eq = 0;
else
    p0_eq = -1*p0*log2(p0);
end
currentEntropy = p1_eq + p0_eq;
```

```
gains = -1*ones(1,numberAttributes);
for i=1:numberAttributes;
    if (activeAttributes(i))
        s0 = 0; s0_and_true = 0;
        s1 = 0; s1_and_true = 0;
        for j=1:numberExamples;
            if (examples(j,i));
                s1 = s1 + 1;
                if (examples(j, numberAttributes + 1));
                    s1_and_true = s1_and_true + 1;
                end
            else
                s0 = s0 + 1;
                if (examples(j, numberAttributes + 1));
                    s0_and_true = s0_and_true + 1;
                end
            end
        end
        if (~s1);
            p1 = 0;
        else
            p1 = (s1_and_true / s1);
        end
        if (p1 == 0);
            p1_eq = 0;
        else
            p1_eq = -1*(p1)*log2(p1);
        end
        if (~s1);
            p0 = 0;
        else
            p0 = ((s1 - s1_and_true) / s1);
        end
        if (p0 == 0);
            p0_eq = 0;
        else
            p0_eq = -1*(p0)*log2(p0);
        end
        entropy_s1 = p1_eq + p0_eq;
        if (~s0);
            p1 = 0;
        else
            p1 = (s0_and_true / s0);
        end
        if (p1 == 0);
```

5.2 Decision Trees

```
                p1_eq = 0;
            else
                p1_eq = -1*(p1)*log2(p1);
            end
            if (~s0);
                p0 = 0;
            else
                p0 = ((s0 - s0_and_true) / s0);
            end
            if (p0 == 0);
                p0_eq = 0;
            else
                p0_eq = -1*(p0)*log2(p0);
            end
            entropy_s0 = p1_eq + p0_eq;
            gains(i) = currentEntropy - ((s1/numberExamples)*entropy_
s1) - ((s0/numberExamples)*entropy_s0);
        end
end
[~, bestAttribute] = max(gains);
tree.value = attributes{bestAttribute};
activeAttributes(bestAttribute) = 0;
examples_0= examples(examples(:,bestAttribute)==0,:);
examples_1= examples(examples(:,bestAttribute)==1,:);
if (isempty(examples_0));
    leaf = struct('value', 'null', 'left', 'null', 'right',
'null');
    if (lastColumnSum >= numberExamples / 2); % for matrix
examples
        leaf.value = 'true';
    else
        leaf.value = 'false';
    end
    tree.left = leaf;
else
    tree.left = id3(examples_0, attributes, activeAttributes);
end
if (isempty(examples_1));
    leaf = struct('value', 'null', 'left', 'null', 'right',
'null');
    if (lastColumnSum >= numberExamples / 2);
        leaf.value = 'true';
    else
        leaf.value = 'false';
    end
```

```
        tree.right = leaf;
else
        tree.right = id3(examples_1, attributes, activeAttributes);
end
return
End
function [nodeids_,nodevalue_] = print_tree(tree)
global nodeid nodeids nodevalue;
nodeids(1)=0; % The value of the root node is 0
nodeid=0;
nodevalue={};
if isempty(tree)
    disp(' null tree!');
    return ;
end
queue = queue_push([],tree);
while ~isempty(queue)
      [node,queue] = queue_pop(queue);
      visit(node,queue_curr_size(queue));
      if ~strcmp(node.left,'null')
          queue = queue_push(queue,node.left);
      end
      if ~strcmp(node.right,'null')
          queue = queue_push(queue,node.right);
      end
end
nodeids_=nodeids;
nodevalue_=nodevalue;
end
function visit(node,length_)
    global nodeid nodeids nodevalue;
    if isleaf(node)
        nodeid=nodeid+1;
        fprintf(' Leaf node, node: %d\t, attribute value: %s\n', ...
        nodeid, node.value);
        nodevalue{1,nodeid}=node.value;
    else
        nodeid=nodeid+1;
        nodeids(nodeid+length_+1)=nodeid;
        nodeids(nodeid+length_+2)=nodeid;
        fprintf('node: %d\t attribute value: %s\t, the left subtree is node:node%d, the right subtree is node:node%d\n', ...
        nodeid, node.value,nodeid+length_+1,nodeid+length_+2);
        nodevalue{1,nodeid}=node.value;
```

```
        end
    end
end
function flag = isleaf(node)
    if strcmp(node.left,'null') && strcmp(node.right,'null')
        flag =1;
    else
        flag=0;
    end
end
```

The results are shown in Fig. 5.19.

5.3 Random Forest

Section 5.2 studied the relevant content of decision trees. This section will study random forests (Random Forest, RF) in detail. The main contents of this section include:

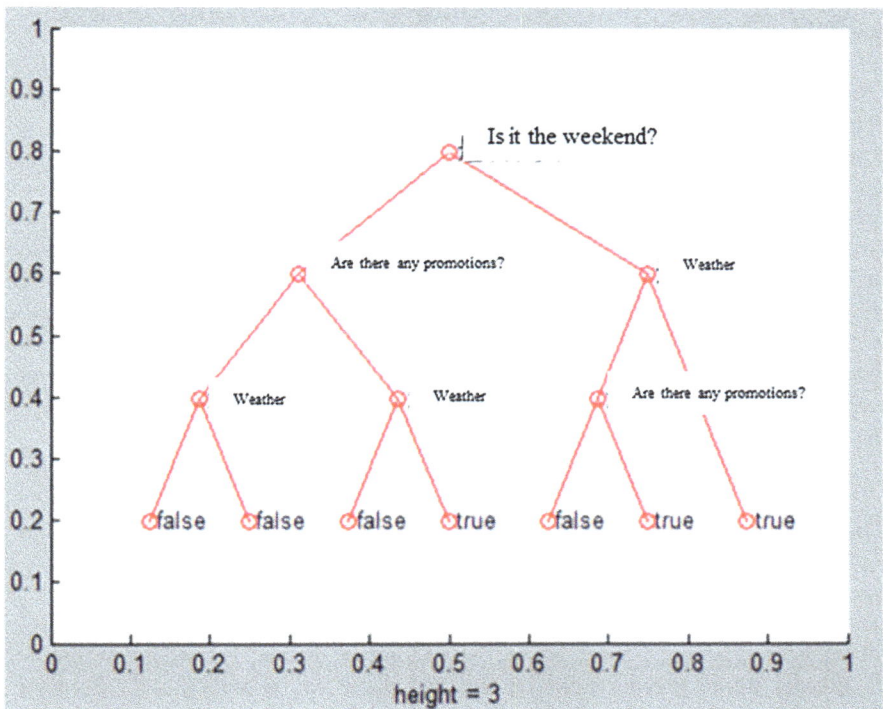

Fig. 5.19 The result of MATLAB program simulation

- The definition and classification principles of random forests;
- The convergence of random forests;
- The characteristics of random forests;
- The construction method of random forests;
- The promotion of random forests.

5.3.1 The Basic Concept of Random Forest

From the introduction of decision tree algorithms, it can be found that decision trees have their own advantages and disadvantages.

1. Classification rules are complex.
2. Convergence to local optima rather than global optima.
3. Prone to over-fitting.

Therefore, many scholars improve prediction accuracy by combining multiple models. These methods are called ensemble or classifier combination methods. The combination method first uses training data to build a group of base-class models, and then makes predictions for each base-class model (because variables are classified) or takes average values (because variables are continuous numerical values) to determine the final prediction value.

To generate these combined models, it is usually necessary to generate random vectors to control the growth of each decision tree in the model. Bagging is an early combination method, also known as bootstrap aggregating, which is a method of generating decision trees by randomly sampling partial samples from the training set (not necessarily with replacement). Another method is random splitting selection, which is to randomly select a split at each node from the k best splits. Breiman proposed a method of randomly sampling output values from the original training set to obtain a new training set. Ho proposed a method for random subspaces, which has been studied a lot. This method selects subsets of feature variables through the method to generate each decision tree. Amit and Geman defined many geometric properties and found the best split for each selected attribute from these random selections. This method was proposed by Breiman in 1996 and has greatly inspired the development of random forests.

The Definition and Classification Principle of Random Forest

The above-mentioned common feature is that the k-th decision tree generates a random vector θ_k, and θ_k is independently and identically distributed with respect to the previous random vector $\theta_1, \theta_2, \cdots, \theta_{k-1}$. Using the training set and random vector θ_k, a decision tree is generated. The obtained classification model is $h(X, \theta_k)$ where X is the input variable (i.e., the independent variable). In the bagging method, the random vector θ_k can be understood as the result of throwing a flying dart at N

boxes. Here, N is the number of records in the training set. After generating many decision trees, the final result is obtained by voting or taking average values, which is called the random forest method (Lu and Song 2016).

The components of a random forest are k decision trees. Each tree's judgment principle for decision-making has no great difference from a single-tree decision-making method, but each tree is not exactly the same. From the original training sample set N, samples are repeatedly and randomly sampled k times to generate a new training sample set. Then, according to the self-help sample set, k classification trees are generated to form a random forest. The classification results of new data are determined by how many votes each classification tree casts. In essence, this is an improvement of the decision-tree algorithm. By combining multiple decision trees together, each tree's dependence on an independent sample is reduced. In the forest, each tree has a similar distribution, and classification errors depend on each tree's classification ability and the correlation between them. Feature selection uses random methods to split each node, and then compares the errors generated under different conditions. It can detect the internal estimation error, classification ability, and related attributes to determine the selection of features. Although the classification ability of a single tree may be small, after generating a large number of decision trees, a test sample can be classified by each tree, and the most likely classification is selected through statistical methods, as shown in Fig. 5.20.

Convergence of Random Forests

Given a set of classification models $\{h_1(X), h_2(X), \cdots, h_k(X)\}$, each classification-based training set is randomly sampled from the original data set (X, Y). Then, the margin function can be obtained:

$$mg(X,Y) = av_k I\left[h_k(X) = Y\right] - \max_{j \neq k} av_k I\left[h_k(X) = j\right]$$

The margin function is used to measure the degree to which the average correct classification data exceeds the average incorrect classification data. The larger the margin value, the more reliable the classification prediction. The generalization error (generalization error) can be written as:

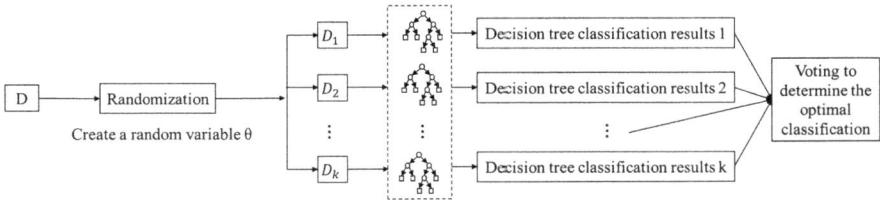

Fig. 5.20 Schematic diagram of a random forest structure

$$PE^* = P_{X,Y}\left[mg(X,Y) < 0\right]$$

When the number of decision-tree classification models is sufficient, $h_k(X) = h_k(X_{\theta k})$ obeys the law of large numbers.

Theorem 5.1 It is proved that as the number of decision-tree classification models increases, all sequences $\theta_1, \theta_2, \cdots, \theta_k$, PE^* almost surely converges to

$$P_{X,Y}\left\{P_\theta\left[h(X,\theta) = Y\right] - \max_{j \neq k} P_\theta\left[h(X,\theta) = j\right] < 0\right\}$$

Proof It can be proved that for all X, in the sequence space $\theta_1, \theta_2, \cdots, \theta_k$, there exists a zero-probability set C (i.e., outside the set C) such that

$$\frac{1}{N}\sum_{n=1}^{N} I\left[h(\theta_n, X) = j\right] \to P_\theta\left[h(\theta_n, X) = j\right]$$

For a given training set and given parameters θ, all sets X that satisfy $h(\theta_n, X) = j$ are hyper-rectangles. For all $h(\theta, X)$, there exists a finite number K of hyper-rectangles, denoted as $S_1 S_2 \cdots S_k$. If $\{X : h(X, \theta) = j\} = S_k$, define $\varphi(\theta_n = k)$. Let be the number of times in the first experiments. Then:

$$\frac{1}{N}\sum_{n=1}^{N} I\left[h(\theta_n, X) = j\right] = \frac{1}{N}\sum_{k} N_k I(X \in S_k)$$

By the law of large numbers, we have:

$$N_k = \frac{1}{N}\sum_{n=1}^{N} I\left[\varphi(\theta_n) = k\right]$$

It converges almost everywhere to $P[\varphi(\theta_n) = k]$. For all unions of sets, given a zero-probability set C (i.e., outside the set C), for all k, convergence does not necessarily exist everywhere.

$$\frac{1}{N}\sum_{n=1}^{N} I\left[h(\theta_n, X) = j\right] \to \sum_{k} P_\theta\left[\varphi(\theta_n) = k\right] I(X \in S_k)$$

The right-hand side is $P_\theta[h(\theta_n, X) = j]$, so it is proved.

This theorem explains why the RFC method does not produce over-fitting problems with the increase of decision trees, but it should be noted that there may be a generalization error within a certain limit.

Characteristics of Random Forests

Random forests have the following advantages:

1. Among the existing algorithms, the accuracy of the random forest algorithm is incomparable.
2. Random forests can efficiently process large-scale data sets.
3. Random forests can handle thousands of input attributes.
4. In classification applications, random forests can calculate the importance of different variable attributes.
5. In the process of constructing a random forest, an internal unbiased estimate of the generalization error can be produced.
6. When a large amount of data is missing, random forests use efficient methods to estimate the missing data and maintain accuracy.
7. In unbalanced data sets, random forests can provide methods to balance errors.
8. Generated random forests can be saved for later use.
9. The calculation of prototypes (Prototypes) can give the correlation between attribute variables and classification.
10. Calculating the proximity (Proximity) between sample instances can be used for cluster analysis, anomaly analysis, or other interesting views of data.

Random forests have the following disadvantages:

1. On some sample sets with large noise, random forest models are prone to over-fitting.
2. Features with more partition values are likely to have a greater impact on the decision-making of random forests, thus affecting the performance of the fitted model.

5.3.2 Construction Method of Random Forests

The specific steps for constructing a random forest are as follows (Liu 2017):

Step 1: If there are N samples, randomly select N samples with replacement (randomly select one sample each time, and then return to continue selection). Use the selected N samples to train a decision tree as the sample at the root node of the decision tree.

Step 2: When each sample has M attributes, randomly select m attributes from these M attributes at each node of the decision tree when splitting is needed, satisfying the condition $m \ll M$. Then, use some strategy (such as information gain) to select one attribute from these m attributes as the splitting attribute for that node.

Step 3: Each node in the decision tree formation process needs to be split according to Step 2 (it is easy to understand that if the attribute selected by the node next time is the same as the attribute used when its parent node was split, then this node has reached a leaf node and does not need to continue splitting) until it is no longer possible to split. Note that there is no pruning in the entire decision tree formation process.

Step 4: According to Steps 1–3, a large number of decision trees are established, thus forming a random forest.

Figure 5.21 is a schematic diagram of repeated sampling.

From the steps above, it can be seen that the randomness of the random forest is reflected in the randomness of the training samples for each tree and the random selection of classification attributes for each node in the tree. With these two random guarantees, the random forest will not produce over-fitting phenomena. The random forest has two parameters that need to be controlled by humans. One is the number of trees in the forest, and it is generally recommended to take a large value; the other is the size of m, and it is recommended that the value of m be the square root of M.

If random sampling is not performed, the training sets for each tree are the same, and thus the tree classification results trained in the end are completely the same.

If the sampling is not with replacement, the training samples for each tree are different and not intersected. That is, each tree is "biased" and is absolutely "one-sided" (of course, this may not be correct), meaning that the training results of each tree are very different. And the final classification of the random forest depends on the voting of multiple trees (weak classifiers), and this voting should be "seeking consensus". Therefore, using completely different training sets to train each tree will not help the final classification result, and this is no different from "blind people touching an elephant".

The classification effect (error rate) of the random forest is related to two factors:

1. The correlation between any two trees in the forest: the greater the correlation, the higher the error rate.
2. The classification ability of each tree in the forest: the stronger the classification ability of each tree, the lower the error rate of the entire forest.

Reducing the number of selected features m will also reduce the correlation and classification ability of the tree; increasing m will also increase both. Therefore,

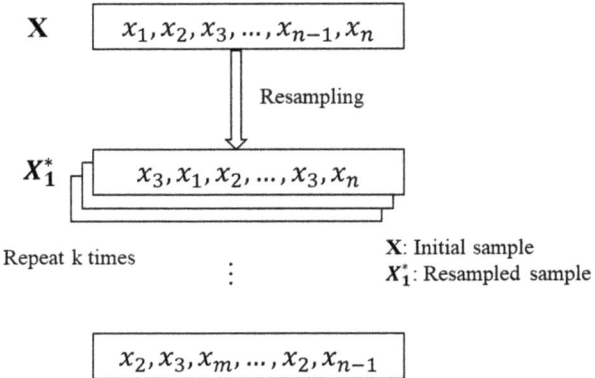

Fig. 5.21 Bootstrap resampling diagram

5.3 Random Forest

the key problem is how to select the optimal m (or range), which is also the only parameter of the random forest.

The key problem in constructing a random forest is how to select the optimal m, and to solve this problem mainly depends on calculating the out-of-bag (OOB) error rate.

Random forests have an important advantage: there is no need to perform cross-validation or use an independent test set to obtain an unbiased estimate of the error. It can be evaluated internally, meaning that an unbiased estimate of the error can be established during the generation process.

When constructing each tree, different bootstrap samples (randomly and with replacement) are used for the training set. So for each tree (assuming for the k-th tree), about 1/3 of the training instances do not participate in the generation of the k-th tree, and they are called the OOB samples of the k-th tree.

This sampling characteristic allows us to perform OOB estimation, and its calculation method is as follows:

1. For each sample, calculate the classification of trees for which it is an OOB sample (about 1/3 of the trees).
2. Use simple majority voting as the classification result for this sample.
3. Use the ratio of the number of misclassified samples to the total number of samples as the OOB error rate of the random forest.

5.3.3 Extension of Random Forests

ET

ET (Extra Trees) is a variation of random forests. The principle is almost the same as that of random forests, with only the following differences:

1. For the training set of each decision tree, random forests use random sampling (bootstrap) to select sample sets as the training set for each decision tree, while ET generally does not use random sampling, that is, each decision tree uses the original training set.
2. After selecting the partition features, the decision tree of the random forest will select an optimal feature value partition point based on criteria such as the Gini coefficient or mean square error, which is the same as the traditional decision tree. However, ET is more aggressive and will randomly select a feature value to partition the decision tree.

Taking a binary tree as an example, when the feature attribute is in the form of categories, random selection will take some samples of these categories as the left branch, and take samples of other categories as the right branch; when the feature attribute is in the form of numerical values, random selection will choose any number between the maximum and minimum values of the feature attribute. When the sample's feature attribute value is greater than this value, it is taken as the left branch, and when it is less than this value, it is taken as the right branch. In this way, the purpose of distributing samples to the two branches based on this feature attribute is achieved. Then calculate the split value at this time (if the feature attribute is in the form of categories, the Gini index can be used; if the feature attribute is in the form of numerical values, the mean square error can be used). Traverse all feature attributes at the node, and obtain the split values of all feature attributes by the above method. The split value with the largest selection value is used to implement the split of this node. It can be seen from the above introduction that this method is more random than the randomness of random forests.

Since the partition points of feature values are randomly selected rather than optimal points, this will generally lead to the size of the generated decision tree being larger than that generated by the random forest. That is to say, the variance of the model is further reduced compared to the random forest, and in some cases, the generalization ability of ET is better than that of the random forest.

For a certain decision tree, because its optimal split feature is randomly selected, its prediction results are often inaccurate when using it. However, when multiple decision trees are combined, good prediction results can be achieved.

After the ET is constructed, the prediction error of the ET can also be obtained by using all the training samples. This is because although the construction of the decision tree and the prediction application use the same training sample set, since the optimal split feature is randomly selected, different prediction results will still be obtained. Comparing these prediction results with the true responses of the samples can obtain the prediction error. Compared with random forests, in ET, all training samples are OOB samples, so calculating the prediction error of ET is equivalent to calculating the OOB error.

Here, only the differences between the ET algorithm and random forests are introduced. The other contents of the ET algorithm (such as the calculation of prediction and OOB errors) are completely the same as those of random forests. For specific details, refer to the introduction of random forests.

TRTE

TRTE (Totally Random Trees Embedding) is a non-supervised data transformation method. It maps low-dimensional data sets to high-dimensional ones, allowing the mapped high-dimensional data to be better used for classification and regression models. In support vector machines, the kernel method is used to map low-dimensional data sets to high-dimensional ones. Here, TRTE provides another method.

TRTE also uses a method similar to RF in the data transformation process. It builds T decision trees to fit the data. After the decision trees are established, the position of each data in the data set at the leaf nodes of the T decision trees is determined. If there are 3 decision trees, and each decision tree has 5 leaf nodes, and a certain data feature x is divided into the second leaf node of the first decision tree, the third leaf node of the second decision tree, and the fifth leaf node of the third decision tree, then the feature of x after mapping is encoded as $(0,1,0,0,0,0,0,1,0,0,0,0,0,0,1)$, which has a 15-dimensional high-dimensional feature. Spaces are added between these feature dimensions to distinguish the encodings of the 3 decision trees.

After mapping to high-dimensional features, various supervised learning classification and regression algorithms can be continuously used.

IForest

IForest (Isolation Forest) is a method for outlier detection. It also uses a method similar to random forests to detect outliers.

IForest Algorithm Principle

IForest belongs to non-parametric and unsupervised methods, neither defining mathematical models nor requiring labeled training. To find which points are likely to be isolated, IForest uses a very efficient strategy. Assume that using a random hyper-plane to cut the data space, each cut can generate two sub-spaces. Continue to use a random hyper-plane to cut each sub-space until there is only one data point left in each sub-space. Intuitively, it can be found that clusters with high densities will be cut many times before stopping, while points with low densities will stop at a sub-space very early.

The IForest algorithm benefits from the idea of random forests. Like random forests composed of a large number of decision trees, the IForest forest is also composed of a large number of binary trees. The trees in IForest are called ITrees (Isolation Trees). ITrees are different from decision trees. Their construction process is also simpler than decision trees and is a completely random process.

Suppose there are N records in the dataset When constructing an ITree, n samples are uniformly sampled (usually without replacement) from the N records as the training samples for this tree. In the sample, a feature is randomly selected, and a value is randomly selected within the range of all values of this feature (between the minimum and maximum values). The sample is bisected according to this value, with samples less than this value divided to the left of the node and samples greater than or equal to this value divided to the right of the node. Thus, a splitting condition and the data sets on the left and right sides are obtained, and then the above process is repeated on the data sets on the left and right sides until the data set contains only one data point.

There is one record or the specified tree height is reached.

Since the abnormal data is small and the feature values are very different from the normal data, when constructing the ITree, the abnormal data is closer to the root, while the normal data is farther from the root. The result of a single ITree is often unreliable. The IForest algorithm constructs multiple binary trees through multiple samplings. Finally, the results of all trees are integrated, and the average depth is taken as the final output depth, thus calculating the abnormal branches of the data points.

Isolation Forest Algorithm Steps

How to cut this data space is the core idea of the IForest design. Here, only the most basic method is introduced. Since the cutting is random, an ensemble method (Monte Carlo method) is needed to obtain a convergence value, that is, starting from the beginning repeatedly, and then averaging the results of each cut. The IForest is composed of t ITrees, and each ITree is a binary tree structure. Therefore, the construction of the ITree is first introduced below, and then the construction of the IForest tree is considered.

Construction of ITree

ITree is a kind of random binary tree, and each node has either two children (i.e., leaf nodes) or no children. Given a data set D, where all attributes of D are continuous variables, the construction process of ITree is as follows:

1. Randomly select an attribute Attr.
2. Randomly select a value Value of this attribute.
3. Classify each record according to Attr, put the records with Attr less than Value on the left child, and put the records with Attr greater than or equal to Value on the right child.
4. Then recursively construct the left child and the right child until the following conditions are met: there is only one record in the input data set or multiple records are the same; the height of the tree reaches the specified height.

After the ITree is constructed, data prediction can be carried out. The prediction process is to put the test records on the ITree and see which leaf nodes the test records are on. The assumption that ITree can effectively detect anomalies is that: abnormal points are generally rare, and in ITree, they will be quickly divided into leaf nodes. Note that abnormal points are generally sparse, so smaller divisions can be used to group them into separate regions, or in other words, the space containing them is relatively large.

Therefore, the length of the path $h(x)$ from the leaf node to the root node can be used to judge whether a record x is an abnormal point (that is, judge whether x is an abnormal point according to $h(x)$). For a data set containing n records, the minimum height of the constructed tree is $log(n)$, and the maximum value is n − 1. Using $log(n)$ and n − 1 for normalization cannot ensure fairness and is inconvenient, so a more complex normalization formula is used:

5.3 Random Forest

$$S(x,n) = 2^{-\frac{h(x)}{c(n)}}$$

$$c(n) = 2H(n-1) - \left[2(n-1)/n\right]$$

$$H(k) = \ln(k) + \xi$$

Among them, $S(x, n)$ is the abnormal index of ITree constructed by the training data of record x in n samples, the value range of $S(x, n)$ is [0, 1], and $\xi = 0.5772156649$.

The record x is the anomaly index of the ITree constructed by the training data of n samples, and the value range is [0, 1]. The judgment of abnormal situations is divided into the following cases:

1. The closer to 1, the higher the possibility of being an outlier.
2. The closer to 0, the higher the possibility of being a normal point.
3. If most of the training samples' $S(x, n)$ are close to 0.5, it indicates that the entire data set has no obvious abnormal values.

If attributes and attribute values are randomly selected, a single tree selected in this random manner is definitely not good. However, when multiple trees are combined, it becomes powerful.

Construction of IForest

Given a data set D containing n records, the method of constructing IForest is somewhat similar to that of random forests. Both randomly sample a part of the data set to construct a tree to ensure the difference between different trees. However, iForest is different from random forests. The amount of sampled data P_{si} does not need to be equal to n, and it can be much smaller than n.

Figure 5.22a is the original data, and Fig. 5.22b is the sampled data. The blue ones are normal samples, and the red ones are abnormal samples. It can be seen that before sampling, normal samples and abnormal samples overlap, so it is difficult to distinguish them. However, after sampling, abnormal samples and normal samples can be clearly separated.

The steps for constructing an IForest are as follows:

1. Randomly select n point-samples from the training data as subsamples and put them into the root nodes of the tree.
2. Randomly specify a dimension (attribute) and randomly generate a cut-point p within the current node data. The cut-point is generated between the maximum and minimum values of the specified dimension within the current node data.
3. Using this cut-point, a hyper-plane is generated, and the current node data space is divided into two sub-spaces: the data with the specified dimension less than p

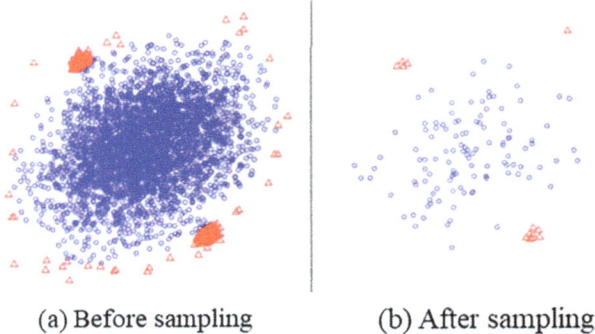

(a) Before sampling   (b) After sampling

Fig. 5.22 Schematic diagram before and after sampling

is placed on the left child of the current node, and the data with the dimension greater than or equal to p is placed on the right child of the current node.

4. Recursively perform steps 2 and 3 on the child nodes until there is only one data point in the leaf nodes (no further cutting is possible) or the leaf nodes have reached the specified height.

In addition to limiting the subsample size, a maximum height should also be set for each ITree. This is because abnormal data records are relatively rare, and their path lengths are also relatively short. Therefore, it is only necessary to distinguish normal records from abnormal records, focusing on parts with lower-than-average heights. This way, the algorithm efficiency is higher. After such adjustments, the calculation needs to be improved bit by bit.

After obtaining t ITrees, the iForest training is completed, and then the iForest is used to evaluate the test data. For a training data X, let it traverse each ITree, and then calculate the height at which X finally falls on each tree (the height of X on the tree). The average path length of X on each tree (average path length over the ITree) can be obtained.

IForest Characteristics

1. IForest has linear time complexity because it uses the method of random forests, so it can be used for data sets with large amounts of data. Generally, the more trees there are, the more stable the algorithm. Since each tree is generated independently of each other, it can be deployed on large-scale distributed systems for accelerated operation.
2. IForest is only sensitive to global sparse points (global anomaly) and is not good at handling local relative sparse points (local anomaly). In some local situations, the detection of anomaly points may not be very accurate. At the same time, IForest is not suitable for very high-dimensional data. Because each time the data space is cut, a dimension and a random feature of that dimension are ran-

domly selected. After the tree is built, there are still a large number of dimensions that are not used, resulting in a decrease in algorithm reliability.
3. IForest is not suitable for very high-dimensional data. Because each time the data space is cut, a dimension is randomly selected. After the tree is built, there are still a large number of dimensions that are not used, resulting in a decrease in algorithm reliability. High-dimensional space may also have a large number of noise dimensions or irrelevant attributes, affecting the construction of the tree.
4. IForest promotes the mass estimation theory. Currently, it has achieved significant results in classification clustering and anomaly detection.
5. The IForest algorithm mainly has two parameters: the number of binary trees; the number of samples drawn when training a single ITree. Experiments show that when the number of binary trees is set to 100 and the number of sampling samples is 256, IForest can achieve good results in most cases, which also reflects the simple and efficient nature of the algorithm.
6. IForest is an unsupervised detection algorithm and is currently one of the most commonly used algorithms for outlier detection. In practical applications, there is no need for black-and-white labels. It should be noted that if the proportion of abnormal samples in the training samples is relatively high, it violates the basic assumptions of abnormal detection mentioned earlier and may affect the final results. Abnormal detection is closely related to specific application scenarios. The "abnormal" detected by the algorithm may not necessarily be what is actually desired. Therefore, when selecting features, it is necessary to filter out less relevant features to avoid identifying some irrelevant "abnormal".

5.3.4 Relevant Applications of Random Forests and MATLAB Examples

Application Example 1

```
clear;clc;close all
   load imports-85;
   Y = X(:,1);
   X = X(:,2:end);
   isCategorical = [zeros(15,1);ones(size(X,2)-15,1)]; %
Categorical variable flag
   tic
   leaf = 5;
   ntrees = 200;
   fboot = 1;
   disp('Training the tree bagger')
```

```
   b = TreeBagger(ntrees, X,Y, 'Method','regression',
'oobvarimp','on', 'surrogate', 'on',
'minleaf',leaf,'FBoot',fboot);
   toc
   disp('Estimate Output using tree bagger')
   x = Y;
   y = predict(b, X);
   toc
   cct=corrcoef(x,y);
   cct=cct(2,1);
   disp('Create a scatter Diagram')
   plot(x,x,'LineWidth',3);
   hold on
   scatter(x,y,'filled');
   hold off
   grid on
   set(gca,'FontSize',18)
   xlabel('Actual','FontSize',25)
   ylabel('Estimated','FontSize',25)
   title(['Training Dataset, R^2=' num2str(cct^2,2)],'Fon
tSize',30)
   drawnow
   fn='ScatterDiagram';
   fnpng=[fn,'.png'];
   print('-dpng',fnpng);
   tic
   disp('Sorting importance into descending order')
   weights=b.OOBPermutedVarDeltaError;
   [B,iranked] = sort(weights,'descend');
   toc
   disp(['Plotting a horizontal bar graph of sorted labeled
weights.'])
   figure
   barh(weights(iranked),'g');
   xlabel('Variable Importance','FontSize',30,'Interpreter'
,'latex');
   ylabel('Variable Rank','FontSize',30,'Interpreter','latex');
   title(...
       ['Relative Importance of Inputs in estimating
Redshift'],...
       'FontSize',17,'Interpreter','latex'...
       );
   hold on
```

5.3 Random Forest

```
    barh(weights(iranked(1:10)),'y');
    barh(weights(iranked(1:5)),'r');
    grid on
    xt = get(gca,'XTick');
    xt_spacing=unique(diff(xt));
    xt_spacing=xt_spacing(1);
    yt = get(gca,'YTick');
    ylim([0.25 length(weights)+0.75]);
    xl=xlim;
    xlim([0 2.5*max(weights)]);
    for ii=1:length(weights)
        text(...
            max([0 weights(iranked(ii))+0.02*max(weights)]),ii,...
            ['Column ' num2str(iranked(ii))],'Interpreter','latex'
,'FontSize',11);
    end
    set(gca,'FontSize',16)
    set(gca,'XTick',0:2*xt_spacing:1.1*max(xl));
    set(gca,'YTick',yt);
    set(gca,'TickDir','out');
    set(gca, 'ydir', 'reverse' )
    set(gca,'LineWidth',2);
    drawnow
    fn='RelativeImportanceInputs';
    fnpng=[fn,'.png'];
    print('-dpng',fnpng);
    disp('Ploting out of bag error versus the number of
grown trees')
    figure
    plot(b.oobError,'LineWidth',2);
    xlabel('Number of Trees','FontSize',30)
    ylabel('Out of Bag Error','FontSize',30)
    title('Out of Bag Error','FontSize',30)
    set(gca,'FontSize',16)
    set(gca,'LineWidth',2);
    grid on
    drawnow
    fn='EroorAsFunctionOfForestSize';
    fnpng=[fn,'.png'];
    print('-dpng',fnpng);
```

The results are shown in Fig. 5.23.

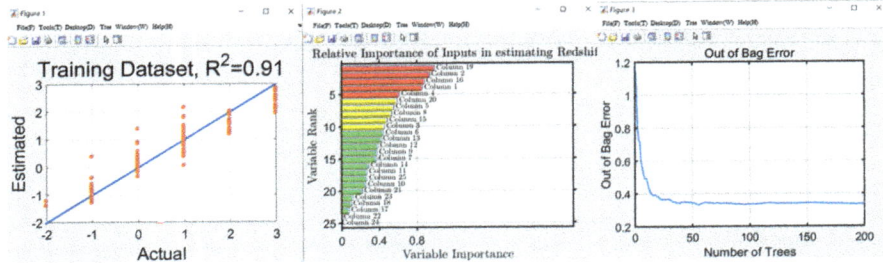

Fig. 5.23 The result graph of MATLAB code implementation

Application Example 1

```
clear all;
   rnode=cell(3,1);
   sn=300;
   tn=10;
   load('aaa.mat');
   n = size(r,1);
   discrete_dim = [];
   for j=1:tn
       Sample_num=randi([1,n],1,sn);
       SData=r(Sample_num,:);
       [tree,discrete_dim]= train_C4_5(SData, 5, 10,
discrete_dim);
       rnode{j,1}=tree;
   end
   load('aaa.mat');
   T = r;
   TData = roundn(T,-1);
   result = statistics(tn, rnode, TData, discrete_dim);
   gd = T(:,end);
   len = length(gd);
   count = sum(result==gd);
   fprintf(' There are %d samples in total, and the number of
correct judgments is %d\n',len,count);
   function [tree,discrete_dim] = train_C4_5(S, inc_node, Nu,
discrete_dim)
       train_patterns = S(:,1:end-1)';
       train_targets = S(:,end)';
       [Ni, M] = size(train_patterns);
       inc_node= inc_node*M/100;
       if isempty(discrete_dim)
   corresponding dimension on the test patterns
```

5.3 Random Forest

```
                discrete_dim = zeros(1,Ni);
                for i = 1:Ni
                    Ub = unique(train_patterns(i,:));
                    Nb = length(Ub);
                    if (Nb <= Nu)
                    end
                end
            end
    flag = [];
    tree    = make_tree(train_patterns, train_targets, inc_node,
discrete_dim, max(discrete_dim), 0, flag);
    function tree = make_tree(patterns, targets, inc_node,
discrete_dim, maxNbin, base, flag)
    [N_all, L]= size(patterns);
        if isempty(flag)
            N_choose = randi([1,N_all],1,0.5*sqrt(N_all));
            Ni_choose = length(N_choose);
            flag.N_choose = N_choose;
            flag.Ni_choose = Ni_choose;
        else
            N_choose = flag.N_choose;
            Ni_choose = flag.Ni_choose;
        end
        Uc = unique(targets);
        tree.dim = 0;
        tree.split_loc= inf;
        if isempty(patterns)
            return
        end
           if ((inc_node > L) | (L == 1) | (length(Uc) == 1))
              H = hist(targets, length(Uc));
              [m, largest] = max(H);
              tree.Nf = [];
              tree.split_loc   = [];
              tree.child = Uc(largest);
              return
           end
           for i = 1:length(Uc)
               Pnode(i) = length(find(targets == Uc(i))) / L;
           end
    log2(9/14) - 5/14 * log2(5/14) = 0.940
        Inode = -sum(Pnode.*log(Pnode)/log(2));
            delta_Ib    = zeros(1, Ni_choose);
            split_loc   = ones(1, Ni_choose)*inf;
        for i_idx = 1:Ni_choose
```

```
            i = N_choose(i_idx);
            data     = patterns(i,:);
            Ud       = unique(data);
            Nbins    = length(Ud);
            if (discrete_dim(i))
                P    = zeros(length(Uc), Nbins);
                for j = 1:length(Uc)
                    for k = 1:Nbins
                        indices = find((targets == Uc(j)) &
(patterns(i,:) == Ud(k)));
                        P(j,k)   = length(indices);
                    end
                end
                Pk= sum(P);
                P1= repmat(Pk, length(Uc), 1);
                P1= P1 + eps*(P1==0);
                P= P./P1;
                Pk= Pk/sum(Pk);
                info= sum(-P.*log(eps+P)/log(2));
                delta_Ib(i_idx) = (Inode-sum(Pk.*info))/
(-sum(Pk.*log(eps+Pk)/log(2)));
            else
                P = zeros(length(Uc), 2);
                [sorted_data, indices] = sort(data);
                sorted_targets = targets(indices);
                  I = zeros(1,Nbins);
                  delta_Ib_inter    = zeros(1, Nbins);
                  for j = 1:Nbins-1
                    P(:, 1) = hist(sorted_targets(find(sorted_
data <= Ud(j))) , Uc);
                    P(:, 2) = hist(sorted_targets(find(sorted_
data > Ud(j))) , Uc);
                    Ps = sum(P)/L;
                    P  = P/L;
                    Pk = sum(P);
                    P1 = repmat(Pk, length(Uc), 1);
                    P1 = P1 + eps*(P1==0);
                    info= sum(-P./P1.*log(eps+P./P1)/log(2));
                    I(j) = Inode - sum(info.*Ps);
                    delta_Ib_inter(j) =   I(j)/
(-sum(Ps.*log(eps+Ps)/log(2)));
                  end
                [~, s] = max(I);
                delta_Ib(i_idx) = delta_Ib_inter(s);
                split_loc(i_idx) = Ud(s);
```

5.3 Random Forest

```
                end
         end
   [m, dim]     = max(delta_Ib);
    dims        = 1:Ni_choose;
     dim_all = 1:N_all;
     dim_to_all = N_choose(dim);
     tree.dim = dim_to_all;
     Nf       = unique(patterns(dim_to_all,:));
     Nbins    = length(Nf);
     tree.Nf = Nf;
     tree.split_loc = split_loc(dim);
     if (Nbins == 1)
         H = hist(targets, length(Uc));
         [m, largest] = max(H);
         tree.Nf = [];
         tree.split_loc  = [];
         tree.child = Uc(largest);
         return
     end
     if (discrete_dim(dim_to_all))
         for i = 1:Nbins
             indices = find(patterns(dim_to_all, :) == Nf(i));
             tree.child(i)= make_tree(patterns(dim_all,
indices), targets(indices), inc_node, discrete_dim(dim_all),
maxNbin, base, flag);
         else

indices1 = find(patterns(dim_to_all,:) <= split_loc(dim));

indices2 = find(patterns(dim_to_all,:) > split_loc(dim));
             if ~(isempty(indices1) | isempty(indices2))
                 tree.child(1)= make_tree(patterns(dim_all,
indices1), targets(indices1), inc_node, discrete_dim(dim_all),
maxNbin, base+1, flag);
                 tree.child(2)= make_tree(patterns(dim_all,
indices2), targets(indices2), inc_node, discrete_dim(dim_all),
maxNbin, base+1, flag);
             else
                 H = hist(targets, length(Uc));
                 [m, largest] = max(H);
                 tree.child = Uc(largest);
                 tree.dim = 0;
             end
     end
   function [result] = statistics(tn, rnode, PValue,
```

```
discrete_dim)
      TypeName = {'1','2'};
      TypeNum = [0 0];
      test_patterns = PValue(:,1:end-1)';
      class_num = length(TypeNum);
      type = zeros(tn,size(test_patterns,2));
      for i = 1:tn
          type(tn,:) = vote_C4_5(test_patterns, 1:size(test_
patterns,2), rnode{i,1}, discrete_dim, class_num);
      end
      result = mode(type,1)';
   end
   function targets = vote_C4_5(patterns, indices, tree,
discrete_dim, Uc)
      targets = zeros(1, size(patterns,2));
      if (tree.dim == 0)
          %Reached the end of the tree
          targets(indices) = tree.child;
          return
      end
      dim = tree.dim;
      dims= 1:size(patterns,1);
      if (discrete_dim(dim) == 0)
          in= indices(find(patterns(dim, indices) <= tree.
split_loc));
          targets= targets + vote_C4_5(patterns(dims, :), in,
tree.child(1), discrete_dim(dims), Uc);
          in= indices(find(patterns(dim, indices) >   tree.
split_loc));
          targets= targets + vote_C4_5(patterns(dims, :), in,
tree.child(2), discrete_dim(dims), Uc);
      else
          Uf = unique(patterns(dim,:));
          for i = 1:length(Uf)
              if any(Uf(i) == tree.Nf)
                  in = indices(find(patterns(dim, indices) ==
Uf(i)));
                  targets = targets + vote_C4_5(patterns(dims,
:), in, tree.child(find(Uf(i)==tree.Nf)), discrete_dim(dims), Uc);
              end
          end
      end
```

The results are shown in Fig. 5.24.

Fig. 5.24 The accuracy rate of algorithm sample data judgment

```
464 samples in total, 427 of them are judged correctly
>> 427/464

ans =

0.9203
```

References

W. He and L. Zhang, Principles and Practices of Machine Learning, Posts & Telecom Press, 2021.
F. Liu, Algorithms in the Era of Big Data: Machine Learning, Artificial Intelligence, and Typical Cases, Electronics Industry Press, Beijing, 2017.
X. Lu and J. Song, Big Data Mining and Statistical Machine Learning, China Renmin University Press, 2016.
B. Ratner, Statistical Mining and Machine Learning, China Machine Press, 2021.
W. Wang, Principles and Applications of Artificial Intelligence, 4th ed., Publishing House of Electronics Industry, 2018.
Z. Wei, Machine Learning, Publishing House of Electronics Industry, 2018.

GPSR Compliance
The European Union's (EU) General Product Safety Regulation (GPSR) is a set of rules that requires consumer products to be safe and our obligations to ensure this.

If you have any concerns about our products, you can contact us on

ProductSafety@springernature.com

In case Publisher is established outside the EU, the EU authorized representative is:

Springer Nature Customer Service Center GmbH
Europaplatz 3
69115 Heidelberg, Germany

www.ingramcontent.com/pod-product-compliance
Ingram Content Group UK Ltd.
Pitfield, Milton Keynes, MK11 3LW, UK
UKHW022203230426
470311UK00001BA/13